No. 260

FISH

A Practical Approach

Edited by

Barbara Beatty

Department of Pathology, University of Vermont
College of Medicine, Burlington, VT, USA

Sabine Mai

Manitoba Institute of Cell Biology,
Winnipeg, Manitoba, Canada

and

Jeremy Squire

Princess Margaret Hospital, Toronto, Ontario
Canada

OXFORD
UNIVERSITY PRESS

OXFORD
UNIVERSITY PRESS

Great Clarendon Street, Oxford OX2 6DP

Oxford University Press is a department of the University of Oxford.
It furthers the University's objective of excellence in research, scholarship,
and education by publishing worldwide in

Oxford New York

Auckland Bangkok Buenos Aires Cape Town Chennai Dar es Salaam
Delhi Hong Kong Istanbul Karachi Kolkata Kuala Lumpur Madrid
Melbourne Mexico City Mumbai Nairobi São Paulo Shangai Taipei
Tokyo Toronto

with an associated company in Berlin

Oxford is a registered trade mark of Oxford University Press in the UK and
in certain other countries

Published in the United States by Oxford University Press Inc., New York

A catalogue record for this title is available from
the British Library

Library of Congress Cataloguing-in-Publication Data
(Data available)

ISBN 0 19 963883 7 (Hbk)
      0 19 963884 5 (Pbk)

10 9 8 7 6 5 4 3 2 1

Typeset in Swift by Footnote Graphics, Warminster, Wilts
Printed in Great Britain on acid-free paper
by The Bath Press, Bath

FISH

The Practical Approach Series

Related **Practical Approach** Series Titles

Essential Molecular Biology V2 2/e
Radioisotopes 2/e
Functional Genomics
Essential Molecular Biology V1 2/e
Differential Display
Mouse Genetics and Transgenics
Gene Targeting 2/e
DNA Microarray Technology
Protein Expression
Transcription Factors 2/e
Chromosome Structural Analysis
In Situ Hybridization 2/e
Chromatin
PCR3: PCR In Situ Hybridization
Antisense Technology
Genome Mapping
DNA Cloning 4: Mammalian Systems
DNA Cloning 3: Complex Genomes
Gene Probes 1
Gene Probes 2
DNA Cloning 1: Core Techniques

Please see the **Practical Approach** series website at

http://www.oup.com/pas

for full contents lists of all Practical Approach titles.

Preface

This book is intended to provide essential practical information for clinical, basic, and student researchers interested in developing and using FISH methods to study genomic and molecular changes in cells. It is structured to not only provide protocols, but also illustrate the great variety of FISH applications with specific examples taken wherever possible from research and clinical experiments being carried out in the contributor's laboratories. We have attempted to provide basic principles, easy to follow protocols, and troubleshooting sections to cover all aspects of FISH from labelling of probes to the use of multicolour FISH on cDNA microarrays and DNA chips. The human emphasis is supplemented with data from other model organisms such as the mouse, and relevant website sources have been included in most chapters.

The stimulus for preparing this book came from the realisation that a current, affordable book addressing the many practical aspects of FISH technology at the appropriate level did not exist. There are other volumes now available but either their expense precludes widespread use, and/or the variety of FISH methods commonly used in research and clinical practice are not covered. The offering of information in this text has been designed to cater to the wide range of abilities among the students, technologists and laboratory researchers and will hopefully lead to more adventurous experiments using this extraordinarily colourful technology!

We wish to thank those who generously contributed the excellent chapters and to all those who have helped to make this book possible. We hope you find this book useful and we welcome comments and suggestions for future editions.

February 2002   B. B.
                                                                                                       S. M.
                                                                                                       J. S.

Contents

List of protocols   *page   xiii*
Abbreviations   *xvii*

1 Introduction   *1*
Sabine Mai, Barbara G. Beatty, and Jeremy A. Squire
   References   *4*

2 FISH probes and labelling techniques   *5*
Patricia Bray-Ward

1 Introduction   *5*
2 Fluorescence principles   *5*
   Fluors and haptens   *6*
   Commonly used fluors   *7*
   Choice of filter sets   *7*
3 Nucleic acid probes   *8*
   Types of probes   *8*
   Preparation of cloned probes   *9*
   Enzymatically amplified probes   *15*
   Synthetic oligonucleotide probes   *16*
4 Coupling of fluors/haptens to nucleotides   *17*
5 Labelling of probes   *19*
   Nick translation   *20*
   Random primer labelling   *21*
   RNA transcription labelling   *22*
   PCR labelling   *24*
6 Post-labelling DNA processing and purification   *25*
   DNase treatment (for FISH and other hybridization protocols)   *25*
   Removal of unincorporated nucleotides and BSA from the reaction mix prior to probe precipitation   *25*
7 Other labelling systems   *26*
   Coupling of fluors or haptens to amine-modified nucleic acids   *26*
   Other chemical coupling systems   *26*
   Direct chemical coupling of fluors or haptens to proteins   *26*
   References   *27*

3 Human chromosome mapping of single copy genes   29

Barbara G. Beatty and Stephen W. Scherer

  1 Introduction   *29*
  2 DNA probes for FISH mapping   *30*
      Identification of FISH probes from WWW sites   *30*
      Preparation of probes for FISH mapping   *32*
  3 Target DNA preparation   *34*
      Metaphase chromosomes   *34*
      Mapping with interphase nuclei   *36*
      Hypotonic treatment and fixation   *38*
  4 Slide preparation   *39*
      Target slide pre-treatment   *41*
  5 Denaturation and hybridization of probe and target DNA   *42*
  6 Post-hybridization washes   *44*
  7 Immunodetection   *45*
  8 Chromosome counterstaining and banding   *47*
  9 Microscopy and image analysis   *48*
 10 FISH mapping points to consider   *49*
      FISH mapping of single probes to metaphase chromosomes   *49*
      Relational mapping with multiple probes   *51*

      References   *53*

4 Murine chromosome preparation   55

Sabine Mai and Francis Wiener

  1 Murine chromosome preparation for banding and *in situ* hybridization
      procedures   *55*
      Introduction   *55*
  2 Giemsa–trypsin banding of mouse chromosomes   *63*
  3 Molecular cytogenetic approaches for murine chromosomes   *66*
      References   *76*

5 High resolution FISH mapping using chromatin and DNA fibre   77

Henry H. Q. Heng

  1 Introduction   *77*
  2 Practical considerations for fibre preparation   *78*
  3 General equipment required for fibre FISH   *78*
  4 Chromatin fibre preparation   *79*
  5 DNA fibre preparation   *83*
  6 FISH   *85*
      DNA probe labelling   *85*
      Hybridization   *86*
      Wash   *88*
      Detection and amplification   *89*
      Counterstaining and antifade   *90*
  7 Photography   *91*
      Acknowledgements   *91*
      References   *91*

6 Applications of RNA FISH for visualizing gene expression and nuclear architecture *93*

Rose Tam, Lindsay S. Shopland, Carol V. Johnson, John A. McNeil, and Jeanne B. Lawrence

1 Introduction *93*

2 Cell preparation *96*
 Detergent-extracted cell preparation *96*
 Cytogenetic preparations *97*

3 Probe preparation *99*

4 Hybridization to RNA *100*
 Basic RNA hybridization procedure *101*
 Oligonucleotide hybridization *102*

5 Hybridization to DNA *103*
 Detecting heat denatured cellular DNA *104*
 DNA FISH using NaOH denaturation and RNA hydrolysis *105*

6 Multiple label techniques and applications *105*
 Coupling the detection of RNA with DNA *105*
 Coupling protein detection with FISH *107*
 Chromosome paints and RNA FISH *108*
 Differentiating transcripts with intron and cDNA probes *109*
 Exon suppression hybridization: an example of the use of specific competition *110*

7 Visualizing and analysing results *112*
 Microscopy *113*
 Digital imaging *114*

8 Concluding remarks *116*

 Acknowledgements *117*

 References *117*

7 FISH on three-dimensionally preserved nuclei *119*

I. Solovei, J. Walter, M. Cremer, F. Habermann, L. Schermelleh, and T. Cremer

1 Introduction *119*

2 Preparation and fixation of cells *123*
 Preparation of slides *123*
 Cultivation and fixation of adherent cells *125*
 Preparation, attachment, and fixation of cells growing in suspension *126*

3 Preparation of cells directly isolated from peripheral blood *127*

4 Pre-treatments needed for hybridization *128*

5 Hybridization set-up *131*
 Probe labelling *131*
 Probe preparation *132*
 DNA denaturation and hybridization *132*

6 Post-hybridization washes, detection, nuclei counterstaining, and slide mounting *134*
 Post-hybridization washes *134*
 Detection of hybridized probes *135*
 Counterstaining of nuclei and mounting cells in antifade medium *137*

7 Combined 3D FISH and replication labelling *139*
 Replication labelling *139*
 Detection of incorporated halogenated deoxyuridines after FISH *140*

CONTENTS

8 Combined protein immunodetection and 3D FISH   142

9 Preservation of the chromatin structure during 3D FISH   144

10 Confocal microscopy   144
   Selection of the filter configuration   147
   Conditions of image acquisition   149
   Calibration of the instrument   150
   Visualization   150
   Quantitative measurements and deconvolution   152

   References   154

8 Comparative genomic hybridization on metaphase chromosomes and DNA chips   159

Stefan Joos, Carsten Schwänen, and Peter Lichter

1 Introduction   159

2 Preparation of metaphase chromosomes   161

3 Isolation of genomic DNA   161

4 Isolation of single cells by micromanipulation   167

5 Amplification of genomic DNA from small cell populations by universal polymerase chain reaction (PCR)   168

6 Probe labelling   172

7 Comparative genomic hybridization   174
   Denaturation of metaphase chromosomes   174
   Probe mixture   175
   In situ hybridization and signal detection   176

8 Image acquisition and evaluation   177

9 Troubleshooting of CGH experiments   178

10 Troubleshooting of CGH experiments in combination with universal PCR   179

11 New developments: matrix-CGH   180

   References   181

9 FISH in clinical cytogenetics   183

Jeremy A. Squire, P. Marrano, and E. Kolomietz

1 Introduction   183

2 Probes commonly used for FISH in the clinical laboratory   184

3 Preparation of clinical samples for FISH analysis   185
   Preparation of metaphase chromosomes for FISH   185
   Preparation of interphase nuclei derived from clinical specimens for FISH   187

4 Criteria for assessing and reporting FISH results   194
   General considerations when selecting cells for FISH microscopy   195
   Scoring criteria for interphase FISH signal evaluation and enumeration   196
   Special considerations concerning interphase FISH interpretation   197

5 Some of the commonly used FISH probes in clinical cytogenetics   198
   FISH analysis of microdeletion syndromes   198
   Use of the three-colour fusion (translocation/inversion) probes   199
   Use of FISH probes in assessing gene amplification   201

6 Appendix (useful web sites for molecular cytogenetics clinical sources)   202

   References   202

10 Multicolour FISH and spectral karyotyping *205*

Jane Bayani and Jeremy A. Squire

1 Introduction *205*

2 Spectral karyotyping (SKY) *206*

3 M-FISH *208*

4 M-FISH and SKY protocols *208*

5 General considerations for image acquisition and analysis *214*
   Image analysis using SKY *215*
   Image analysis using M-FISH *216*

6 Troubleshooting *216*
   References *219*

11 cDNA microarrays for fluorescent hybridization analysis of gene expression *221*

Javed Khan, Lao H. Saal, Michael L. Bittner, Yuan Jiang, Gerald C. Gooden, Arthur A. Glatfelter, and Paul S. Meltzer

1 Introduction *221*
   Serial analysis of gene expression *221*
   Oligonucleotide arrays *222*
   cDNA microarrays *222*

2 Fabrication of cDNA microarrays *223*
   Culturing cDNA bacterial clones *223*
   Microarray slide printing *230*

3 Target production *232*

4 Hybridization *235*

5 Image acquisition *237*

6 Image analysis and normalization *237*

7 Sensitivity and specificity *238*

8 Data mining and statistical analysis *238*

9 Summary *239*
   References *239*

List of suppliers *241*

Index *251*

Protocol list

Nucleic acid probes

Alkaline lysis preparation of plasmid DNA   *9*

Preparation of phage DNA   *10*

DNA isolation from cosmids, PAC clones, and BAC clones   *12*

Preparation of yeast genomic DNA   *14*

Creating DOP-PCR probe libraries   *15*

Coupling of fluors/haptens to nucleotides

Coupling of fluor or hapten to dUTP   *18*

Labelling of probes

Nick translation   *20*

Random primer labelling   *21*

RNA transcription labelling   *23*

PCR labelling   *24*

Other labelling systems

Succinimidyl ester labelling of proteins   *26*

DNA probes for FISH mapping

Ethanol precipitation of labelled probe DNA   *33*

Target DNA preparation

Culture and harvest of peripheral blood lymphocytes   *35*

Cell synchronization and BrdU incorporation   *36*

Preparation of fibroblast G0–G1 interphase nuclei   *37*

Hypotonic treatment and cell fixation   *38*

Slide preparation

Slide preparation   *40*

Target slide pre-treatment   *41*

Denaturation and hybridization of probe and target DNA

DNA denaturation and hybridization   *43*

Post-hybridization washes

Post-hybridization washes   *44*

Immunodetection

Detection of probes labelled with biotin or DIG   *46*

Murine chromosome preparation for banding and *in situ* hybridization procedures

Chromosome preparations from mouse bone marrow cells   *57*

Pellet fixation   *58*

Suspension fixation   *59*

Alternative suspension fixation   *59*

Chromosome preparation from mouse splenic cells   *60*

Chromosome preparation from mouse thymic (and lymph node) cells   *60*

Chromosome preparation of mouse plasmacytoma   *61*

Preparation of mouse chromosome spreads for molecular cytogenetics   *62*

Determination of the mitotic index of cell populations used for chromosome preparations   *62*

Giemsa–trypsin banding of mouse chromosomes

Giemsa–trypsin banding of mouse chromosomes   *64*

Ageing the slides by H_2O_2 pre-treatment   *65*

Molecular cytogenetic approaches for murine chromosomes

Labelling of probes for FISH, assessment of labelling efficiency, and titration of probes   *68*

Determination of labelling quality (spot test)   *69*

Fluorescent *in situ* hybridization   *70*

Detection of hybridization following FISH   *71*

Chromosome painting   *72*

Spectral karyotyping (SKY)   *74*

Banding after SKY   *76*

Chromatin fibre preparation

Preparation of chromatin fibre from cultured lymphocytes by drug treatment   *80*

Preparation of chromatin fibres with alkaline buffer   *81*

Preparation of chromatin fibre using cytospin   *82*

DNA fibre preparation

DNA fibre preparation   *83*

DNA fibre preparation using gel blocks   *84*

FISH

Probe labelling   *85*

Probe denaturation and pre-hybridization   *86*

Slide preparation for hybridization   *87*

Hybridization   *88*

Wash   *88*

Detection with signal amplification   *89*

Counterstaining and antifade   *90*

Cell preparation

Extraction of monolayer cells   *97*

Preparation of cytogenetic cells and chromosomes   *98*

Probe preparation

Nick translation of DNA probes   *100*

Hybridization to RNA

Basic RNA hybridization procedure   *101*

Hybridization to DNA
Heat denaturation of DNA   *104*
Alkaline denaturation of DNA   *105*

Multiple label techniques and applications
Sequential detection of RNA and DNA   *106*
Immunostaining for proteins   *107*
Hybridization with chromosome paint   *109*
Exon suppression   *111*

Preparation and fixation of cells
Preparation of slides   *124*
Subculture and fixation of adherent cells on slides   *125*
Preparation and fixation of cells growing in suspension   *126*

Preparation of cells directly isolated from peripheral blood
Isolation of mononuclear cells directly from peripheral blood   *127*

Pre-treatments needed for hybridization
Post-fixation treatments   *130*

Hybridization set-up
Probe loading, denaturation, and hybridization   *133*

Post-hybridization washes, detection, nuclei counterstaining, and slide mounting
Post-hybridization washes   *134*
Detection of haptens in hybridized probes   *136*
Counterstaining of nuclei and mounting in antifade medium   *138*

Combined 3D FISH and replication labelling
Replication labelling by the incorporation of halogenated deoxyuridines   *139*
Detection of incorporated halogenated nucleotides after 3D FISH   *141*

Confocal microscopy
Preparing slides with fluorescent beads   *151*
Measurement of the chromatic shift   *151*

Preparation of metaphase chromosomes
Preparation of metaphase chromosomes from peripheral blood cells   *164*

Isolation of genomic DNA
Isolation of genomic DNA from blood   *165*
Isolation of genomic DNA from solid tissue samples   *166*
Isolation of genomic DNA from paraffin-embedded tissue   *167*

Isolation of single cells by micromanipulation
Micromanipulation of single cells   *168*

Amplification of genomic DNA from small cell populations by universal polymerase chain reaction (PCR)
Amplification of small amounts of genomic DNA by DOP-PCR   *169*
Amplification of genomic DNA from single cells by SCOMP-PCR   *170*

Probe labelling
Labelling of probe DNA by nick translation   *172*

Comparative genomic hybridization

Denaturation of chromosomal DNA on slides   *174*

Probe mixture and denaturation   *175*

In situ hybridization   *176*

Preparation of clinical samples for FISH analysis

Performing FISH on metaphase spreads   *188*

Preparation of paraffin sections for FISH   *191*

Isolation of intact nuclei from paraffin-embedded tissue for FISH   *193*

Criteria for assessing and reporting FISH results

Pre-screening evaluation and determination of hybridization efficiency   *194*

Selection criteria for FISH analytical microscopy   *195*

General guidelines for performing interphase FISH assays   *196*

M-FISH and SKY protocols

Probe labelling   *208*

Spectral karyotyping metaphase spread pre-treatment and denaturation   *209*

M-FISH metaphase spread pre-treatment and denaturation   *210*

Spectral karyotyping probe precipitation and hybridization to pre-treated denatured slides   *211*

M-FISH probe precipitation and hybridization to pre-treated denatured slides   *212*

Spectral karyotyping post-hybridization washes and hapten detection   *213*

M-FISH post-hybridization washes and hapten detection   *214*

Fabrication of cDNA microarrays

Culturing clones for arrays   *225*

Isolation of plasmid DNA   *226*

PCR amplification of clones   *226*

Quantification of PCR product   *227*

PCR product purification   *228*

Poly-L-lysine pre-treatment of glass slides   *229*

Blocking slides after printing with succinic anhydride   *231*

Target production

RNA extraction   *232*

Direct labelling of cDNA with fluorescent dyes   *234*

Target purification   *234*

Hybridization

Microarray hybridization   *235*

Post-hybridization slide washes   *236*

Abbreviations

BAC	bacterial artificial chromosomes
BrdU	5-bromodeoxyuridine
BSA	bovine serum albumin
CCD	charged-couple device
CGH	comparative genomic hybridization
CLSM	confocal laser scanning microscopy
DIG	digoxigenin
DMSO	dimethyl sulfoxide
DNP	dinitrophenyl
DOP	degenerate oligonucleotide primed
EST	expressed sequence tag
FCS	fetal calf serum
FISH	fluorescence *in situ* hybridization
FITC	fluorescein isothiocyanate
HSR	homogeneously stained region
M-FISH	multicolour FISH
PAC	P1 artificial chromosomes
PBS	phosphate-buffered saline
PCR	polymerase chain reaction
PFA	paraformaldehyde
PHA	phytohaemagglutinin
PI	propidium iodide
PMT	photomultiplier tube
PNA	peptide nucleic acid
PSF	point spread function
RBC	red blood cell
RI	refractive index
SAGE	serial analysis of gene expression
SCOMP	single cell comparative genomic hybridization
SKY	spectral karyotyping
SSC	saline sodium citrate
YAC	yeast artificial chromosomes

Chapter 1
Introduction

Sabine Mai
Manitoba Institute of Cell Biology, 675 McDermot Avenue, Room 6046,
Winnipeg R3E 0V9, Canada.

Barbara G. Beatty
Department of Pathology, University of Vermont College of Medicine,
Room 206 Health Sciences Research Facility, Burlington, VT 05405, USA.

Jeremy A. Squire
Ontario Cancer Institute, Princess Margaret Hospital, 610 University Avenue,
Toronto, Ontario M5G 2M9, Canada.

Human chromosomes have been studied for over a century (1, 2), but not until the 1950s was it determined that the human chromosome complement was diploid and comprised of 46 chromosomes (3, 4). With the development of chromosome pre-treatment procedures and modification of DNA staining techniques in the late 1960s, the identification of each chromosome based on the definition of highly reproducible trypsin–Giemsa or fluorescent banding patterns was obtained. This landmark advance in human cytogenetics allowed researchers to rapidly address specific clinical and research questions and to make excellent progress in understanding the chromosomal basis of many genetic diseases and cancer. To this day classical cytogenetics using banding methods continues to play an important role in the identification and characterization of many different types of chromosomal abnormalities.

Although banding methodologies readily allow the distinction of individual chromosome pairs and the detection of gross karyotype aberrations, the analysis is heavily dependent on the accumulated experience and subjective interpretation of the cytogeneticist. In addition, the analytical phase is very time-consuming and there is a lower limit on the resolving power of conventional cytogenetics for the correct interpretation of more complex karyotypes. Usually the chromosomal origin(s) of unidentified marker chromosomes generates composite banding patterns that cannot be resolved with confidence or precision, so that many cryptic translocations, small insertions, small inversions, small amplifications, or tiny deletions may go undetected. Moreover, all banding technologies depend on good quality cytogenetic preparations of metaphase plates that are well-spread with no chromosome overlaps, and that have been dried and aged appropriately. The low mitotic index, poor growth rate, and suboptimal chromosome morphology, which are often hallmarks of tumour metaphase preparations, make cancer cytogenetics a particularly difficult discipline.

Fluorescence *in situ* hybridization (FISH) analysis of chromosomes began when classical cytogenetics was combined with recombinant DNA technology to form a new discipline called molecular cytogenetics. Prior to FISH the more cumbersome approach had been to perform *in situ* hybridization using radioactive nucleic acid probes for detection of specific DNA or RNA sequences in metaphase chromosomes or interphase (5). However, the use of non-radioactive probes became widely feasible in the early 1980s when methods for labelling nucleic acids with non-radioactive haptens such as biotin became available (6). FISH was actually introduced in the late 1970s (7, 8) but was not used to visualize DNA sequences using human metaphase preparations until 1986 (9). The advantages of FISH over radioactive *in situ* hybridization include increased stability, safer handling, more rapid results, more precise spatial localization, less background, and ability to use multiple colours on one slide making it by far the method of choice for genome and molecular cytogenetic studies today.

The versatility of FISH is exemplified by its successful use in the study of the location of DNA sequences on metaphase chromosomes, in chromatin fibre FISH mapping, DNA microarray quantitation, and in analysis of RNA expression in cell nuclei and cytoplasm. FISH is readily adaptable to a variety of clinical specimen types such as pathology tissue sections and cytological smears. Some of the significant advances in FISH technology across the past two decades have been suppression hybridization (10), simultaneous multicolour FISH (11), fibre FISH (12, 13), comparative genomic hybridization (CGH) (14), and most recently, spectral karyotyping (SKY) (15), M-FISH (16), and FISH/CGH on microarrays (17). In the following chapters, protocols and applications are presented for performing FISH on a number of different substrates in which diverse research and clinical questions can be addressed.

FISH began with the discovery that nucleic acids could be chemically modified to incorporate a hapten such as biotin or digoxigenin, which in turn could be detected with a fluorescently labelled reporter molecule such as avidin or anti-digoxigenin. Since then probe preparation and labelling techniques have been modified and simplified. Now nucleotides can be labelled with fluors directly and incorporated into FISH probes, eliminating the often laborious detection steps. FISH has now become an essential tool for gene mapping and characterization of chromosome aberrations. As the target DNA remains intact, unlike in molecular genetic analysis, information is obtained directly about the positions of probes in relation to chromosome bands (thus 'mapping' the probes), or to other hybridized probes. Chromosome aberrations can therefore be defined using differentially labelled probes, which delineate particular chromosomes or chromosomal regions. The variety of probes—genomic, cDNA, oligonucleotide, and specialized oligonucleotide probes such as molecular beacons and peptide nucleic acid (PNA) probes—and targets make FISH an extremely versatile way to study structure and function of DNA and RNA *in situ*.

High resolution FISH mapping and ordering of probes relative to one another can be performed on released chromatin fibres, and is termed fibre FISH. Strands of chromatin released from the nucleus are allowed to stream or are stretched across a glass slide. Fibre FISH has a wide range of resolution (1 kb–1 Mb) and several methods have been developed for producing and stretching the DNA fibres with or without the presence of histone proteins (12, 13).

One of the major advantages of FISH over conventional molecular biology approaches is that of providing molecular information in the context of cell morphology. Targeting nuclear RNA and the corresponding genes not only within cells but also within a single cell or from a single allele can provide important information about gene expression, processing, and transport of transcripts in normal and mutant cells. The use of RNA FISH for studying the intracellular localization of RNA has increased our understanding of the *in situ* physical characteristics of DNA transcription and transport of RNA transcripts. Similarly, FISH can be used to examine many interesting biological questions about nuclear organization. Three-dimensional nuclear DNA FISH can provide high resolution information about subchromosomal domains, gene position, and the relationship of genes and their transcripts in different cells and during different stages of the cell cycle. Accurate analyses of three-dimensional FISH is highly dependent on excellent quality confocal microscopy and image analysis procedures.

A variation of FISH technology that allows for genome-wide screening of chromosomal gains and losses is comparative *in situ* hybridization (CGH) (14). It is based on the comparison of genomic DNA from two different genomes, and identifies chromosomal gains and losses of one genome relative to the other. CGH is performed on normal chromosome metaphase spreads, which is a distinct advantage for studying tumour samples. The resolution of identifying chromosomal gains and losses on metaphase chromosomes is several Mbs. However, this technique has been modified to increase the resolution to several kbs by the technique of matrix or array CGH (17) in which the targets are cloned DNA fragments immobilized on a glass surface. This allows detection of low copy number gains and losses, and may be used diagnostically to identify microdeletions or amplifications affecting only one or two genes.

Clinical cytogenetics, especially cancer cytogenetics has benefited greatly from FISH technology. FISH often provides essential additional information not available using standard cytogenetic analyses alone. The technique is ideal for clinical laboratories as it is rapid and can be performed on tissue (fresh, frozen, or formalin-fixed paraffin-embedded), touch preps, cytospins, or cell cultures. Since a major difficulty in tumour cytogenetics is the inability to obtain chromosome spreads from many types of tumours, the use of interphase FISH directly on tumour samples (biopsies, sections, and archived paraffin-embedded material) enables the determination of chromosomal aberrations without the need for metaphase chromosome preparations. Numerical chromosome aberrations, chromosome deletions, and translocations can all be identified in interphase nuclei providing important diagnostic and/or prognostic information.

The advent of spectral dyes and imaging has made FISH more colourful and even more powerful. Using multiple probes simultaneously provides important

3

additional information that can now be obtained for a single sample using multi-colour FISH techniques. These techniques allow for both a genome-wide screen of aberrations and a gene- or chromosomal region-specific analyses of specific aberrations in chromosomes and can be adapted for use in the analysis of inter-phase nuclei. Similarly, genome-wide screens for mRNA expression differences or for genomic aberrations can be performed by microarray FISH. Microarray FISH is based on the comparative hybridization of two samples on to arrays that represent either specific sets of genes or the whole genome. The targets used come as oligonucleotide, cDNA, or genomic arrays.

All of the FISH technologies described here rely on excellent fluorescence microscopy with well-optimized image acquisition and analytical software. In addition some of the procedures described require extensive bioinformatics and computational analysis to interpret the results of FISH analysis. Thus we have moved away from subjective analyses of cytogenetic differences between normal and aberrant chromosomes and have entered into a phase of analysis that allows us to examine any possible tissue or chromosome by multiple means using much more objective methods. This book presents an overview of current FISH techniques: general principles, methodology, and applications. It describes the development of this powerful technology and demonstrates its great versa-tility. We hope that this book will help many FISH users and will inspire you to creatively use these tools to develop new avenues of research and diagnostics.

References

1. Arnold, J. (1879). *Virchows Arch. Pathol. Anat.*, **77**, 181.
2. Hansemann, D. V. (1891). *Virchows Arch. Pathol. Anat.*, **123**, 356.
3. Tjio, J. H. and Levan, A. (1956). *Hereditas*, **42**, 1.
4. Hamerton, J. L. (1971). *Human cytogenetics* Vol. II. p. 65. Academic Press, NY.
5. Gall, J. G. and Pardue, M. L. (1969). *Proc. Natl. Acad. Sci. USA*, **64**, 600.
6. Langer, P. R., Waldrop, A. A., and Ward, D. C. (1981). *Proc. Natl. Acad. Sci. USA*, **78**, 6633.
7. Rudkin, G. T. and Stollar, B. D. (1977). *Nature*, **265**, 472.
8. Bauman, J. G., Wiegant, J., Borst, P., and van Duijn, P. (1980). *Exp. Cell Res.*, **128**, 485.
9. Pinkel, D., Straume, T., and Gray, J. (1986). *Proc. Natl. Acad. Sci. USA*, **83**, 2934.
10. Lichter, P., Cremer, T., Borden, J., Manuelidis, L., and Ward, D. C. (1988). *Hum. Genet.*, **80**, 224.
11. Ried, T., Baldini, A., Rand, T. C., and Ward, D. C. (1992). *Proc. Natl. Acad. Sci. USA*, **89**, 1388.
12. Heng, H. H. Q., Squire, J., and Tsui, L-C. (1992). *Proc. Natl. Acad. Sci. USA*, **89**, 9509.
13. Parra, I. and Windle, B. (1993). *Nature Genet.*, **5**, 17.
14. Kallioniemi, A., Kallioniemi, O-P., Sudar, D., Rutovitz, D., Gray, J. W., Waldman, F., *et al.* (1992). *Science*, **258**, 818.
15. Schröck, E., du Manoir, S., Veldman, T., Schoell, B., Weinberg, J., and Ferguson-Smith, M. A. (1996). *Science*, **273**, 494.
16. Speicher, M. R., Gwyn Ballard, S., and Ward, D. C. (1996). *Nature Genet.*, **12**, 368.
17. Solinas-Toldo, S., Lampel, S., Stilgenbauer, S., Nickolenko, J., Benner, A., Dohner, H., *et al.* (1997). *Genes Chromosome Cancer*, **20**, 399.

Chapter 2

FISH probes and labelling techniques

Patricia Bray-Ward

Department of Genetics, Yale University School of Medicine, New Haven, CT 06510, USA.

1 Introduction

Nucleic acid hybridization assays play a central role in both basic research and clinical diagnosis. Until the mid 1980s virtually all probes were labelled isotopically. Although ^{32}P-, ^{125}I-, ^{3}H-, or ^{35}S-labelled DNA or RNA probes provided excellent sensitivity, they had numerous drawbacks, most notably safety issues and short usable shelf-life. Over the past decade methods for labelling probes with haptens, chemiluminescence emitters, and fluorophores have been streamlined and their detection sensitivities improved significantly. This chapter will focus on procedures for the preparation and non-isotopic labelling of probes for applications in fluorescence *in situ* hybridization (FISH).

2 Fluorescence principles

A fluor is a chemical moiety that, in response to absorption of a photon of energy provided by a light source with a suitable emission wavelength, develops an excited electron singlet state. This creates a transient conformational change and partial dissipation of the excitation energy. This is followed by emission of a photon of lower energy and therefore longer wavelength. The difference in energy or wavelength of excitation and emission is called the Stokes shift. This is a critical element in fluorescence imaging since detection systems must exclude wavelengths of light from the excitation source. Stokes shift can be affected by the physical characteristics of the fluor, for example covalent linkage to a protein, the polarity and hydrophobicity of the solvent, and the pH of the solution. Generally fluors with larger Stoke shifts are preferred since they allow filter sets to be used that have wider bandwidths to permit maximum capture sensitivity. Key characteristics of a fluor include its excitation and emission wavelengths, chemical stability, spectral response to pH changes, susceptibility to photobleaching, and quantum yield. Information on these fluor properties usually is provided by commercial sources (Molecular Probes, Amersham Pharmacia

5

Biotech, Roche Diagnostics, etc.). Photobleaching is an important fluor characteristic since the rate of fluor destruction upon illumination influences the imaging times and detection sensitivity. This is particularly important for imaging with mercury or xenon lamps where imaging times are longer relative to laser scanning detection. While the use of antifade reagents can reduce this problem, fluors with greater photostability are generally preferable.

The choice of fluors must be matched to the emission wavelength of the light source that will be used. There are three common light sources—mercury lamps, xenon lamps, and lasers. Mercury and xenon arc lamps produce wide spectrum emissions. Mercury arc lamps provide emission spikes at 254, 366, 436, and 546 nm, matching some absorption peaks of commonly used fluors. Xenon arc lamps provide more even illumination across the 250–1000 nm spectrum. While most commercially available fluors can be used with a mercury lamp, fluors with wavelengths above 700 nm should be avoided. Xenon is sometimes preferred to mercury when more uniform excitation at different wavelengths is desired and when fluors with excitation wavelengths above 700 nm are used. Mercury and xenon arc lamps are used with filter sets that control the wavelength of light that reaches the sample. Usually fluors are excited individually to distinguish their emissions, although double and triple bandpass filters are used in some multiplex imaging applications. Lasers differ by having a very narrow range of emitted light, restricting the fluor choice. Laser-based detection systems frequently have multiple filters for detection of multiple fluors, although special reagents can permit the excitation of multiple fluors simultaneously by fluorescence energy transfer (FRET). Lasers do not require the use of excitation filters. The two commonly used lasers are argon ion lasers, which emit light at 488 and 514 nm, and the helium neon (HeNe) laser, which emits at 543, 594, and 633 nm.

2.1 Fluors and haptens

It is important to research any new fluor before choosing it since fluors differ, not only in excitation, emission, and photostability but also in solubility, sensitivity to solvents, excitation or emission shifts after protein coupling or in different pH solutions. Some fluors do not work well when directly coupled to DNA probes (e.g. Cy3.5 and Cy5.5), presumably because they affect the binding of the probe to target, but are excellent when coupled to secondary detection reagents such as antibodies. The selection of fluors is constantly changing as new fluors are developed. Manufacturers and other researchers can often provide information about the ability of new fluors to be incorporated by polymerases as well as other key characteristics such as photostability.

One can choose to either use probes that have fluors directly attached (direct detection) or probes that have haptens attached that are detected by reagents with attached fluors (indirect detection). Experimental results can be obtained most rapidly and inexpensively when nucleic acid probes are used that have fluors incorporated. Various experimental circumstances may make indirect detection preferable, even though the protocols will be more complicated and more reagents will be needed. Fluors and haptens are frequently incorporated into DNA and RNA probes enzymatically. When a large number of nucleic acid targets are detected in a single experiment, there may not be enough fluors available that can be directly incorporated into probe. In this case, haptens such as biotin, digoxigenin, or oestrogen coupled to dUTP may be enzymatically incorporated and detected with secondary reagents that carry fluor labels. BrdU triphosphate has the advantage of being readily incorporated by any DNA polymerase but can be detected only in single-stranded DNA because of the steric orientation of the bromine residue in double-stranded DNA. In addition, signal amplification can be obtained by using secondary detection or when necessary, antibody sandwiches. The protocol for coupling fluors to avidin or antibody reagents for nucleic acid detection is identical to that used for the chemical coupling of fluor to the antibody directed towards proteins and detection protocols have been developed for simultaneous detection of proteins and nucleic acids.

2.2 Commonly used fluors

The fluors listed in *Table 1* are all photostable enough for most applications and are available coupled to nucleotides, secondary detectors such as antibodies, or as succinimidyl esters (reagents for coupling). Some fluors photobleach much more readily than others, making them virtually unusable. When a very high level of sensitivity is required, necessitating the use of higher wattage light sources or long exposure times, it will be necessary to choose fluors based largely on their photostability. Fluorescein, for example, will photobleach faster than some of the more recently developed dyes such as Alexa488 that have the same excitation and emission characteristics. The behaviour of fluors may differ depending on its conjugation environment, usually exhibited as changes in excitation or emission spectra. Fluors with wavelengths above approximately 650 nm cannot be seen through the microscope with the naked eye and can only be used in conjunction with other fluors or counterstains that can be used for focusing and with a suitable digital camera. For this reason they are rarely used.

2.3 Choice of filter sets

The choice of filter sets is relatively simple but requires a basic understanding of the use of filters in fluorescence detection. Filters are purchased in sets of excitation, dichroic mirror (beam splitter), and emission filters. The light first reaches an excitation filter that allows only the wavelengths of light required for excitation of the fluor to be transmitted. The light then passes through a dichroic mirror (beam splitter) that reflects the light towards the sample. The light emitted by the sample (fluorescence emission) is reflected by the dichroic mirror to the emission filter. The emission filter blocks excitation light reflected off the sample as well as other extraneous light that isn't within the wavelengths permitted to pass through the filter. Excellent filter sets are available from a number of different commercial sources either as off-the-shelf or designed to specifica-

Table 1 Common fluors

	M_r^a	Abs[b]	Em[c]
Fluor			
Diethylaminocoumarin (DEAC)	350	432	472
Fluorescein (FITC)	600	491	515
Rhodamine Green (RGr)	620	515	530
Carboxyrhodamine 6G (R6G)	550	524	552
Cy3 (Cy3)	750	550	570
6-carboxytetramethylrhodamine (TAMRA)	640	547	573
Texas Red (TxR)	800	583	603
Cy5 (Cy5)	800	649	670
Hapten			
Biotin (BIO)	550	–	–
Digoxigenin (DIG)	600	–	–
Dinitrophenyl aminohexanoic acid (DNP)	400	–	–
Bromodeoxyuridine (BrdU)	–		
Fluors suitable for indirect detection			
Oregon Green 488 (OG-488)	510	495	521
Alexa488	650	493	517
Cy3.5	1100	581	596
Cy5.5 (Cy5.5)	1100	675	694
Less suitable fluors include:			
Cascade Blue (CB)	600	396	410
AMCA	450	353	442

[a] M_r is the approximate molecular weight of the dyes or haptens as succinimidyl esters.

[b] Abs = peak absorption (nm).

[c] Em = peak emission (nm).

tion. Since filter sets are designed by experienced individuals that understand the requirements for the emission, beam splitter and excitation filters, the primary consideration for the researcher in choosing filters to match fluors is bandwidth. When experiments are planned that require detection of a large number of individual targets, i.e. the use of many fluors, narrow bandpass filters must be used to insure minimal crossover detection between different fluor windows. When only one or two fluors will be used, wide bandpass filters can be chosen to maximize sensitivity of detection.

3 Nucleic acid probes

3.1 Types of probes

Three general classes of nucleic acid probes can be used, cloned probes, enzymatically amplified probes, and chemically synthesized probes. Cloned probes include: yeast artificial chromosomes (YAC), P1 artificial chromosomes (PAC), and bacterial artificial chromosomes (BAC), cosmid, phage, and plasmid probes. DNA from cloned sources can be first isolated, then propagated enzymatically by making a DOP-PCR library or amplified by isothermal DNA amplification methods. Enzymatically amplified probes can include PCR amplified painting probes for whole genomes, individual chromosomes or sets of chromosomes (or regions), or amplification libraries of cloned probes. In addition, sense and anti-sense RNA probes can be made using RNA polymerase with the T3, T7, or SP6 transcription promoters for RNA analysis. Chemically synthesized probes include simple oligonucleotides and specialized oligonucleotide probes such as molecular beacons or other oligonucleotide probes with various nucleotide analogues incorporated during or after synthesis.

3.2 Preparation of cloned probes

When preparing cloned DNA for FISH probes, usually it is not necessary to eliminate host or vector DNA from the preparation since they neither efficiently hybridize to the mammalian target sequences nor significantly reduce the labelling efficiency of the probes. The need for chemically clean DNA preparations varies with the labelling method. For nick translation labelling, pure DNA is very desirable. For PCR labelling, the cleanliness of the preparation is relatively unimportant. In fact, PCR labelling may be done by simply adding the host containing the clone to the PCR mix prior to the initial thermal denaturation step.

3.2.1 Plasmid preparation

Plasmid DNA usually is labelled by nick translation since the insert sizes required for efficient detection in FISH are larger than the sizes that can be readily amplified by PCR. The standard alkaline lysis preparation of plasmid DNA works well. Plasmid preparation kits also are available from a number of commercial sources (e.g. Qiagen, Roche Diagnostics).

Protocol 1

Alkaline lysis preparation of plasmid DNA[a]

Equipment and reagents

- 15 ml centrifuge tubes
- Clinical centrifuge or equivalent
- 1.5 ml microcentrifuge tubes
- Bacterial media with appropriate antibiotic
- Solution I: 50 mM glucose, 25 mM Tris–HCl, 10 mM EDTA, 5 mg/ml lysozyme pH 8.0; add lysozyme to solution I just before resuspending the bacteria
- Solution II: 0.2 M NaOH, 1% SDS
- Solution III: 5 M potassium acetate (60 ml of 5 M potassium acetate, 11.5 ml of glacial acetic acid, 28.5 ml ddH$_2$O) cooled on ice
- Isopropanol
- 70% ethanol
- 10 mM Tris–HCl, 1 mM EDTA pH 8.0.

Protocol 1 continued

Method

1 Pick a single bacterial colony from a plate and grow overnight in 10 ml of appropriate media.

2 Spin down the culture in a 15 ml centrifuge tube in a clinical centrifuge for 10 min at top speed.

3 Add 200 μl solution I to the centrifuge tube and pipette the bacteria up and down to resuspend. Transfer to an Eppendorf tube. Leave at room temperature for 5 min.

4 Add 400 μl freshly made solution II and mix by inversion. Let stand on ice for 10 min.

5 Add 300 μl ice-cold solution III and leave on ice for 10 min or more. The prep can be left overnight at 4 °C at this point.

6 Centrifuge out the debris at top speed in a microcentrifuge. The debris should form a tight pellet.

7 Remove the supernatant to a fresh microcentrifuge tube. Re-centrifuge if necessary to remove all the visible debris.

8 Precipitate by the addition of 0.6 vol. of isopropanol (approx. 540 μl). Incubate at room temperature for 15 min, then centrifuge in a microcentrifuge at top speed for 20 min at room temperature.

9 Wash the pellet with 70% ethanol and resuspend in 100 μl of 10 mM Tris–HCl, 1 mM EDTA pH 8.0. Plasmid preparations typically yield 1–5 μg/ml.

[a] Derived from Birnboim and Doly (1).

3.2.2 Preparation of phage DNA

Phage DNA is most easily prepared using DNA kits, for example the phage DNA kit from Qiagen, Inc. DNA prepared using *Protocol 2* is clean enough for FISH mapping.

Protocol 2

Preparation of phage DNA

Equipment and reagents

- 50 ml culture flasks
- Vis spectrophotometer
- Bacterial incubator/shaker
- 50 ml polypropylene tubes
- SS34 rotor, RC5B centrifuge, or equivalent
- NZCYM media (Invitrogen)
- Chloroform
- Phenol buffered with 10 mM Tris–HCl, 1 mM EDTA pH 8.0
- DNase (Sigma)
- 500 mM EDTA
- 10% SDS
- Isopropanol
- 10 mM Tris–HCl, 1 mM EDTA pH 8.0

Method

1 Combine 10 ml of NZCYM media with 50 μl of fresh overnight culture of a suitable bacterial host in a 50 ml culture flask.

2 Grow with vigorous shaking for 3–4 h until the culture reaches an OD_{600} of 0.5.

3 Inoculate with 2×10^8 pfu and continue incubation at 37 °C with vigorous shaking for 3–5 h until the culture lyses.

4 Add 200 μl chloroform per flask and incubate for 10 min more.

5 Centrifuge at 1900 g at 10 °C in an SS34 rotor, RC5B centrifuge, or equivalent for 10 min.

6 Transfer the supernatant to a new 50 ml polypropylene tube and warm to room temperature.

7 Add DNase to 1 μg/ml.

8 Incubate for 30 min at room temperature.

9 Add 0.5 ml of 500 mM EDTA and 0.5 ml of 10% SDS.

10 Add 10 ml buffered phenol, mix for 5 min.

11 Add 10 ml of chloroform, mix for 5 min.

12 Centrifuge at 12 000 g at 10 °C in an SS34 rotor, RC5B centrifuge, or equivalent for 10 min.

13 Transfer the supernatant to a new 50 ml tube, re-extract with 10 ml chloroform. Re-centrifuge as in step 12.

14 Transfer the supernatant to a new 50 ml tube, add 10 ml isopropanol, mix well, and allow to sit at room temperature for 10 min.

15 Centrifuge at 17 000 g at 10 °C in an SS34 rotor, RC5B centrifuge, or equivalent for 30 min.

16 Wash the pellet with 70% ethanol.

17 Redissolve in 1 ml of 10 mM Tris–HCl, 1 mM EDTA pH 8.0. The expected yield is 12 μg/10 ml culture.

3.2.3 Cosmid, P1 artificial chromosome (PAC), and bacterial artificial chromosome (BAC) DNA preparations

Cosmid clones carry genomic DNA inserts from 15–30 kb in length, PAC clones (2) contain approx. 100 kb, and BACs (3) approx. 200 kb. DNA probes prepared from these clones provide excellent FISH signals without secondary amplification. However, these DNA preparations give much less DNA than obtained for plasmid preparations since large insert clones are present in bacteria in much lower copy number than plasmids. DNA may be prepared from these clones by a small scale preparation method and the DNA converted to a clone PCR library by DOP-PCR (*Protocol 5*) or by one of several isothermal DNA amplification methods.

Alternatively large scale and higher purity preparations can be made using one of the DNA preparation kits that are commercially available (e.g. Qiagen, Roche Diagnostics). Frequently better FISH results are obtained by using nick translated probes than those labelled by PCR, although nick translation requires larger amounts of more highly purified DNA template.

The preparation of cosmid DNA is influenced by the host bacteria, the vector, the size of insert, as well as other variables such as the bacterial growth conditions prior to DNA preparation. The use of host bacteria derived from HB101 strain, such as TG1 and Top10F, may require increased volumes of buffers throughout the purification procedure to obtain reasonable yields. Cosmid libraries are less frequently used now that PAC and BAC libraries have been developed.

Protocol 3

DNA isolation from cosmids, PAC clones, and BAC clones[a]

Equipment and reagents

- 15 ml snap-cap polypropylene tube
- Shaker
- 37 °C incubator
- Clinical centrifuge or equivalent
- Microcentrifuge
- 1.5 ml microcentrifuge tubes
- LB or TB media
- 25 μg/ml kanamycin (for PAC DNA)
- P1 (filter sterilized, 4 °C): 50 mM glucose, 25 mM Tris–HCl, 10 mM EDTA, 5 mg/ml lysozyme pH 8.0; add lysozyme to P1 just before resuspending the bacteria

- 20 μg/ml chloramphenicol (for BAC DNA)
- TE: 10 mM Tris–HCl pH 8.0, 1 mM EDTA
- 100 μg/ml RNase A
- P2 (filter sterilized, room temp, freshly prepared): 0.2 M NaOH, 1% SDS
- 0.2 M NaOH
- 1% SDS
- P3 (autoclaved, 4 °C): 5 M potassium acetate (60 ml of 5 M potassium acetate, 11.5 ml of glacial acetic acid, 28.5 ml ddH$_2$O)
- 3 M KOAc pH 5.5

Method

1 Inoculate a single isolated bacterial colony into 2 ml TB (or LB) media supplemented with suitable antibiotic (chloramphenicol for CEPH and Sanger Centre BACs, and kanamycin for DeJong RPC1–5 PACs). Use a 15 ml snap-cap polypropylene tube. Grow overnight (up to 16 h) shaking at 225–300 r.p.m. at 37 °C.

2 Pellet the bacteria by centrifuging at top speed in a clinical centrifuge for 15 min.

3 Discard supernatants. Resuspend each pellet in 300 μl P1 solution. Transfer the resuspended bacterial pellets to microcentrifuge tubes. Add 300 μl P2 solution to each tube and gently shake the tubes to mix the contents. Let sit at room temperature for 5 min. The appearance of the suspension should change from very turbid to almost translucent.

Protocol 3 continued

4 Slowly add 300 μl P3 solution to each tube and gently shake during addition. A thick white precipitate of protein and E. coli DNA will form. After adding P3 solution to every tube, place the tubes on ice for 10 min.

5 Place the tubes in a cold microcentrifuge and spin at 8000 g for 10 min.

6 Remove the tubes from the microcentrifuge and place on ice. Transfer supernatants to new microcentrifuge tubes. Avoid transferring any white precipitate material. Re-centrifuge if necessary. Add 800 μl ice-cold isopropanol to each sample. Mix by inverting the tubes a few times; place the tubes on ice for at least 5 min.

7 Spin in cold microcentrifuge for 15 min.

8 Remove the supernatant and wash the DNA pellet by adding 500 μl of 70% EtOH to each tube. Invert the tubes several times and spin in a cold microcentrifuge for 5 min.

9 Remove as much of the supernatant as possible. Occasionally, pellets will become dislodged from the tube so it is better to carefully aspirate off the supernatant rather than pour it off.

10 Air dry the pellets at room temperature. When the DNA pellets turn from white to translucent in appearance, i.e. when most of the ethanol has evaporated, resuspend each in 40 μl TE. Do not use a narrow bore pipette tip to mechanically resuspend DNA sample; rather, allow the solution to sit in the tube with occasional tapping of the bottom of the tube. For large clones resuspension may take over 1 h. The yield from 1 ml of culture is expected to be 0.2–1 μg (cosmids), 0.4 μg (PACs), and 0.1–0.2 μg (BACs).

[a] This is a rapid alkaline lysis miniprep method that is a simplification of the Qiagen preparation method.

3.2.4 Preparation of YAC (yeast artificial chromosome) DNA

YACs are usually 300 kb–1.5 Mb in size, depending on the library and individual clone, with vector sequences contributing only a few kb. However, approx. 50% of the YAC clones in some libraries are chimeric clones, with more than one human DNA fragment being artificially joined during library construction (4). Chimerism may give rise to FISH signals on two or more chromosomes. Sometimes multiple signals also result from cross-hybridization of low copy number repetitive elements in the YAC DNA to other genomic sites. Many YAC clones contain internal rearrangements or deletions, which are usually not detected by FISH. The DNA prepared by the following protocol is mostly yeast genomic DNA; only a small fraction of the DNA is YAC DNA. This is sufficiently pure for FISH. For in situ mapping of YACs, the DNA can be either nick translated or labelled by PCR. If PCR labelling is used, the human DNA insert is often preferentially amplified using inter-Alu PCR primers (5).

Protocol 4

Preparation of yeast genomic DNA

Equipment and reagents

- Clinical centrifuge
- Microcentrifuge
- Water-bath or heat block
- YAC clone in yeast
- AHC medium: mix 6.7 g yeast N2 base (with $(NH_4)_2SO_4$, but without amino acids), 10 g casein hydrolysate (casein amino acids, vitamin grade), H_2O to 800 ml. Adjust to pH 5.8, and autoclave. Allow to cool to RT and add: 100 ml of 20% sterile-filtered dextrose, 4 ml of 5 mg/ml sterile-filtered adenine,[a] 1 ml of 50 mg/ml sterile-filtered ampicillin, 95 ml sterile H_2O, and store at 4°C.
- YPD medium: 15 g yeast extract, 30 g Bacto peptone, H_2O to 800 ml. Autoclave, allow to cool to RT, and add: 1 ml of 50 mg/ml sterile-filtered ampicillin, 100 ml of 20% sterile-filtered dextrose, 99 ml sterile H_2O, and store at 4°C.

- Zymolase 20T (ICN Biomedicals)
- 1 M sorbitol, 0.1 M Na_2EDTA pH 8.0
- 10% SDS
- 50 mM Tris–HCl pH 7.4, 20 mM Na_2EDTA pH 8.0
- 5 M potassium acetate: 60 ml of 5 M potassium acetate, 11.5 ml of glacial acetic acid, 28.5 ml ddH_2O
- Isopropanol
- TE pH 7.4: 10 mM Tris–HCl, 1 mM Na_2EDTA pH 7.4
- TE pH 8.0: 10 mM Tris–HCl, 1 mM Na_2EDTA pH 8.0
- RNase A: 10 mg/ml in 10 mM Tris–HCl pH 7.5, 15 mM NaCl, diluted 1:10 with ddH_2O just before use
- 3 M sodium acetate pH 5.2

Method

1 Pick 2–3 μl cells from a 96-well plate into 1.5 ml AHC in a 50 ml tube. Shake overnight at 30°C. Add 6 ml of 1.5 × YPD and shake overnight at 30°C. (Also okay to pick into 10 ml AHC on Friday for prep on Monday.)

2 Collect the cells in a clinical centrifuge at 500 g for 3 min. Discard the supernatant. The pellet may be stored at 4°C overnight.

3 Resuspend the cells in 0.5 ml of 1 M sorbitol/0.1 M Na_2EDTA pH 8. Transfer to a 1.5 ml microcentrifuge tube.

4 Add 20 μl of Zymolase 20T (12.5 mg/ml of Zymolase 20T, prepared fresh, in 1 M sorbitol/0.1 M Na_2EDTA pH 8) and incubate at 37°C for 60 min.

5 Centrifuge for 1 min.

6 Discard the supernatant. Resuspend (pipette) the cells in 0.5 ml of 50 mM Tris–HCl pH 7.4, 20 mM Na_2EDTA pH 8. Cells will be a little sticky, but must be completely resuspended.

7 Add 50 μl of 10% SDS; mix well (limited vortex okay).

8 Incubate the mixture at 65°C for 30 min.

9 Add 0.2 ml of 5 M potassium acetate, mix thoroughly (limited vortex okay), and place the microcentrifuge tube in ice for 60 min (okay to store at 4 °C overnight).

10 Centrifuge for 5 min.

11 Transfer supernatant to a fresh 1.5 ml tube and repeat centrifugation step.

12 Transfer the supernatant to a fresh 1.5 ml tube and add 0.75 ml of 100% isopropanol at room temperature. Mix by inversion and allow it to sit at room temperature for 5 min. Centrifuge very briefly (10–15 sec). Pour off the supernatant and air dry the pellet, which should be visible.

13 Resuspend the pellet in 0.3 ml of TE pH 7.4.

14 Add 15 μl of 1 mg/ml RNase A and incubate at 37 °C for 30 min.

15 Add 30 μl of 3 M sodium acetate pH 5.2 and 0.2 ml of 100% isopropanol. Mix by inversion and incubate at room temperature for 5 min. Centrifuge briefly (10–15 sec). If you don't see a pellet, spin for 2 min. You should see DNA, but a smaller pellet than in step 12.

16 Pour off the supernatant and air dry. Resuspend the pellet of DNA in 100 μl TE pH 8.0. The yield is usually 5–10 μg of DNA at 50–100 ng/μl when resuspended.

[a] Adenine may be increased to as much as 12 ml—in some cases this improves growth—adjust H_2O volume accordingly.

3.3 Enzymatically amplified probes

3.3.1 PCR amplification of probes and probe libraries

Probe sequences can be amplified from any DNA source with specific primers and may be maintained as an amplifiable fragment or labelled at the time of amplification. Complex probe libraries, such as chromosome painting probes (6) are created and maintained by PCR amplification, usually using the primers described by Telenius and co-workers (7).

Protocol 5

Creating DOP-PCR probe libraries

Equipment and reagents

- Thermocycler
- 10 × PCR buffer: 500 mM KCl, 100 mM Tris–HCl pH 8.3, 15 mM $MgCl_2$
- 25 μM primer: 5′ CCG ACT CGA GNN NNN NAT GTG G 3′
- Taq polymerase (Promega)
- 2 mM dNTP (Pharmacia)
- 0.1–100 ng template DNA, depending on the template complexity

Protocol 5 continued

Method

1 Mix on ice in thin-walled PCR tubes:
- 10 μl of 10 × PCR buffer
- 10 μl primer(s) (20 μM stock, final concentration 2 μM)
- 10 μl dNTPs (2 mM each, final concentration is 200 μM)
- 0.5 μl Taq (5 U/μl stock)
- 1–2 μl of DNA template
- Water to 100 μl

2 Place in the thermocycler and run the following program.
   (a) Denaturation: 1 min at 94°C.
   (b) Low stringency cycles (5 cycles): 94°C for 30–60 sec (denaturation), 30°C for 1.5 min (annealing), 30–72°C for 3 min (ramp), 72°C for 3 min (extension).
   (c) High stringency cycles (35 cycles): 94°C for 30–60 sec (denaturation), 62°C for 1 min (annealing), 72°C for 3 min (extension).

3.3.2 Maintenance of probe libraries

Reamplification of probe libraries is done using the same protocol as the original library creation except that the primers contain only the unique 5′ end sequence of the original primer, 5′ CCG ACT CGA G 3′ and the low stringency amplification cycles are omitted.

3.3.3 Isothermal amplification

A number of isothermal amplification strategies for amplifying DNA have been described in the literature, including strand displacement amplification (SDA) (8), nucleic acid sequence-based replication (NASBA) (9), self-sustaining sequence replication (3SR) (10), and loop-mediated isothermal amplification (LAMP) (11). Kits are becoming available to use some of these technologies for amplification.

3.4 Synthetic oligonucleotide probes

3.4.1 Simple oligonucleotides

The length of oligonucleotide that can be efficiently chemically synthesized is shorter than the optimum length for unique sequence FISH probes. Oligonucleotide probes are useful for detection of high frequency repeat sequences such as centromeres or chromosome bands (12). The use of Alu repeat probes, co-hybridized with unique sequence probes, permits the relative placement of probes on the cytogenetic map (13). The use of repeat hybridization for banding is important in species other than human that do not have clear banding patterns in standard preparations. Oligonucleotides can be chemically synthesized with fluor, hapten, or amino groups incorporated during the synthesis. A limited selection of fluors can be chemically incorporated during oligonucleotide synthesis, primarily because of the sensitivity of many fluors to the reagents used in chemical synthesis. Oligonucleotide synthesis facilities and reagent company web sites are constantly being updated as new fluors and haptens become available. When a specific fluor is required that is not available as a phosphoramidite derivative, then an amine group may be incorporated instead and the fluor is then coupled, post-synthesis. Kits for amine substitution of DNA followed by fluorophore coupling are available commercially (e.g. Molecular Probes). The oligonucleotide probes must be purified before coupling because they are normally contaminated with amines, particularly ammonium, that would interfere with the coupling reaction.

3.4.2 Specialized oligonucleotide probes

A wide variety of oligonucleotide probe designs can be made that incorporate differences in secondary structure, chemistry, or inclusion of quenchers as well as fluors. Probes can be specifically designed to have secondary structures that bring quencher and fluor molecules in close physical proximity. They are designed to change structure on binding to targets such that the quencher and fluor are physically separated resulting in fluorescence in the presence of an appropriate light source. Two examples of such probes are molecular beacons (14) and cyclicons (15). Peptide nucleic acid, PNA, is an artificial construct in which the pentose–phosphate backbone is replaced by polyaminoethyl glycine, a polymer of peptide-like bonds that mimic the ribose–phosphate backbone geometry (16). The use of modified bases in such constructs can be used to increase the melting temperature differential between correctly and incorrectly matched probes. PNA probes offer the additional advantage that decreasing salt concentration causes PNA:DNA to bind more strongly while the opposite is true for DNA:DNA binding.

4 Coupling of fluors/haptens to nucleotides

The decision to buy nucleotides or immunoglobulins pre-labelled, use labelling kits, or do the labelling reactions with lab-made ingredients will depend on the laboratory budget, the volume of reagents needed, the frequency of their use, and the fluor choices. Laboratories using large quantities of labelled reagents often can obtain a sufficient volume discount with commercial suppliers to make purchase of pre-labelled, pre-tested reagents advantageous. Kits for labelling proteins and nucleic acids provide a simple, robust labelling method. This is particularly important when smaller quantities of less commonly used reagents are used, such as a unique antibody. The purchase of pre-labelled reagents makes sense if a labelled reagent is required for single use, in very low quantities, or if a moderate amount of a single common reagent (e.g. fluorescein dUTP) is needed. In addition, some reagents must be purchased as pre-labelled reagents, for example Alexa488 dUTP, or as labelling kits, for example Alexa488 protein labelling kits, when the precursors are not commercially available. However, the high cost of pre-labelled reagents may preclude projects that require substantial amounts of labelled reagent. Labelling nucleotide or immunoglobulin reagents provides an easy, inexpensive alternative to buying pre-labelled reagents or kits.

The protocols for labelling reagents are robust and rapid and the choice of reagent–fluor combinations is limited only by the available fluor/filters. One important consideration is the lability of many labelling reagents. Because of this, lab-made nucleotides should be used only if they are needed constantly in sufficient amounts to produce a fairly rapid turnover in reagents.

Amine-reactive reagents are available in several chemical forms, including succinimidyl esters, isothiocyanates, sulfonyl chlorides. Of these, succinimidyl esters are the chemical form of choice. A wide variety of fluors, including all the fluors listed in the section above, are commercially available as succinimidyl esters. Although it is recommended that fluors be stored as lyophilized solids in a desiccator at –20 °C, they may also be dissolved in dry dimethyl sulfoxide and aliquoted for storage at –20 °C or –80 °C for a period of a few months without significant degradation. The only amine nucleotide that is commercially available is 5-(3-aminoallyl)-2'-deoxyuridine 5'-triphosphate (Sigma Corp.). Various commercial sources sell fluor- or hapten-coupled dCTP.

For those who have large scale FISH requirements, it can be considerably cheaper to synthesize fluor or hapten labelled nucleotides. *Protocol 6* is a straightforward method, that can be performed by individuals not trained in chemistry (17). The products, while not chemically pure, are perfectly adequate for labelling FISH probes for a broad-spectrum of FISH experiments.

Protocol 6

Coupling of fluor or hapten to dUTP

Equipment and reagents

- Microcentrifuge tubes
- Microbalance
- 5-(3-aminoallyl)-2'-deoxyuridine 5'-triphosphate (Sigma Corp.)
- 0.2 M sodium bicarbonate pH 8.3
- Dimethyl sulfoxide (DMSO)
- 2 M glycine pH 8.0
- 1 M Tris–HCl pH 7.5
- One or more of the following fluor or hapten succinimdyl esters (SE) solutions: 20 mM DEAC (7-diethylaminocoumarin-3-carboxylic acid, succinimidyl ester, DEAC-SE, Molecular Probes) in DMSO, 40 mM FITC (fluorescein-5-EX, succinimidyl ester, Molecular Probes) in DMSO, 10 mM TAMRA (6-carboxytetramethyl-rhodamine, succinimidyl ester, 6-TAMRA, Molecular Probes) in DMSO, 40 mM R6G (5-(and-6)-carboxyrhodamine 6G,succinimidyl ester, 5(6)-CR6G, Molecular Probes) in DMSO,

40 mM TxR (Texas RedÆ-X, succinimidyl ester mixed isomers, TxR, Molecular Probes) in DMSO, 20 mM Cy3 (Cyanine3, succinimidyl ester, Cy3 Osu mono, Amersham/Pharmacia) in DMSO, 20 mM Cy3.5 (Cyanine3.5, succinimidyl ester, Cy3.5 Osu mono, Amersham/Pharmacia) in DMSO, 20 mM Cy5 (Cyanine5, succinimidyl ester, Cy5 Osu mono, Amersham/Pharmacia) in DMSO, 40 mM RGr (Rhodamine Green-X, succinimidyl ester, RG-X, Molecular Probes) in DMSO, 40 mM BIO—6((6((biotinoyl)amino) hexanoyl) amino)hexanoic acid, succinimidyl ester (biotin-XX, Molecular Probes) in DMSO, 40 mM DIG—3-amino-3-deoxydigoxigenin hemisuccinamide, succinimidyl ester (digoxigenin, Molecular Probes) in DMSO, or 40 mM DNP—6-(2,4-dinitrophenyl)aminohexanoic acid, succinimidyl ester (DNP-X, Molecular Probes) in DMSO.

Protocol 6 continued

Method

1 Mix the ingredients in the sequence shown from left to right in the table, where v is volume. A convenient labelling protocol uses 1v = 10 μl and yields 1 mM fluor dUTP solution after dilution to 200 μl final volume.

SE[a]	aadUTP[b]	H$_2$O	Buffer	DMSO	mM[c] SE	Vol. SE
DEAC	1v	1v	1v	–	20	1v
TAMRA	1v	1v	1v	1v	10	2v
Cy5	1v	1v	1v	–	20	1v
FITC	1v	1.5v	1v	–	40	0.5v
RGr	1v	1.5v	1v	1v	40	0.5v
R6G	1v	1.5v	1v	1v	40	0.5v
TxR	1v	1.5v	1v	1v	40	0.5v
BIO	1v	1.5v	1v	–	40	0.5v
DIG	1v	1.5v	1v	1v	40	0.5v
DNP	1v	1.5v	1v	1v	40	0.5v

3 Incubate the reaction mixture in the dark for 3–4 h at room temperature.

4 Add 0.2v of 2 M glycine pH 8.0 (20 mM final concentration) and 0.4v of 1 M Tris–HCl pH 7.5 (20 mM final concentration) to stop the reaction and stabilize the nucleotides.

5 Dilute the reactions to 200 μl final volume with 10 mM Tris–HCl, 1 mM EDTA pH 8.0.

[a] SE indicates the name of the fluor or hapten.

[b] aadUTP is 5-(3-aminoallyl)-2′-deoxyuridine 5′-triphosphate, dissolved at 20 mM in reaction buffer (0.2 M sodium bicarbonate pH 8.3).

[c] mM indicates the millimolar concentration at which the reactive dye was dissolved in DMSO.

Nucleotides can be labelled and used in PCR or nick translation reactions without prior purification using *Protocols 8* and *11*. The addition of relatively large amounts of BSA in the reaction mix necessitates the purification of the labelled DNA products (described in Section 6.2) by phenol extraction or proteinase K incubation followed by precipitation prior to addition to hybridization mix when large amounts of probe are used (e.g. for M-FISH where 100 μl of more of probe solution is precipitated).

5 Labelling of probes

Probe labelling usually has the double aim of introducing labelled nucleotides into the probe sequence and reducing the DNA fragment size to under 500 bases in length. Longer probes can cause significant background problems. When very short probes are used (e.g. 45-mers), hybridization and wash conditions are adjusted by lowering the percentage of formamide, increasing the concentration of SSC, or lowering the wash temperatures. As probe size decreases, the length of wash time should also decrease.

5.1 Nick translation

Nick translation is the method of choice when sufficient clean DNA template is available. Nick translation uses the simultaneous activity of two enzymes: DNase I, which creates random nicks in the template and. *E. coli* DNA polymerase I, which through it's 5'-3' exonuclease activity removes nucleotides, while the 5'-3' polymerase activity adds nucleotides to all the available 3' ends created by the DNase.

Protocol 7

Nick translation

Equipment and reagents

- 15°C water-bath (place water-bath in cold room)
- Agarose gel electrophoresis equipment
- 10 × NT buffer: 500 mM Tris–HCl pH 7.5, 100 mM MgCl$_2$, 0.5 mg/ml BSA
- 100 mM 2-mercaptoethanol
- 3 mg/ml DNase I in 50% glycerol, PBS: stored in aliquots at –20°C
- 1 mM d(ACG)TP: 1 mM each, aliquoted in small volumes, and stored at –20°C
- 5 mM dTTP: aliquoted in small volumes and stored at –20°C
- 1 mM labelled dUTP: aliquoted in small volumes and stored at –20°C
- 0.5 μl *E. coli* DNA polymerase I (10 U/μl stock)
- 1–8 μl DNA (final concentration 20–30 ng/μl)
- Stop solutions: 0.5 M EDTA pH 8, and 20% SDS
- 1.5% agarose gel in TAE: 0.04 M Tris–acetate, 1 mM EDTA pH 8.0, with appropriate size markers
- Gel stain solution

Method

1 Prepare fresh 10 × DNase solution by mixing 1 μl of 3 mg/ml DNase stock solution in 1 ml water. Keep on ice.

2 For a 100 μl nick translation reaction, mix:
   - 10 μl of 10 × NT buffer
   - 1.5 μl of 10 μg/ml BSA (if custom nucleotides are being used)
   - 10 μl of 10 × DNase I solution
   - 5 μl dNTP (stock, 1 mM each of dATP, dCTP, dGTP)
   - Water to 100 μl final volume

Protocol 7 continued

3   If using nucleotides prepared by *Protocol 6*, add:
 - 1.7 μl dTTP (1 mM stock)
 - 1.2–8.0 μl of 1 mM custom labelled dUTP
   - 2 μl DEAC (final 20 μM)
   - 3.5 μl R6G, TAMRA, TxR (final 30 μM)
   - 5.5 μl Cy3, Cy5 (final 50 μM)
   - 8 μl AMCA, FITC, BIO, DIG, DNP (final 60–70 μM)

4   If using commercial nucleotides, add:
 - 3.5 μl dTTP (1 mM stock)
 - 0.75–1.75 μl commercial or custom labelled dUTP (1 mM stock)
   - 0.75 μl DEAC
   - 1 μl R6G, TAMRA, TxR
   - 1.25 μl Cy3, Cy5
   - 1.75 μl AMCA, FITC, BIO, DIG, DNP

5   Add:
 - 2.5 μl *E. coli* DNA polymerase I (10 U/μl stock)
 - 5–40 μl DNA (final concentration 20–30 ng/μl)

6   Incubate the reaction for 2 h at 15 °C.

7   Place the reaction on ice while checking the size of labelled fragments (200–500 bp is the desirable size) by TAE agarose gel electrophoresis. Heat denature the samples prior to loading on the gel. If the size is too large, put the reaction back in for an additional 30 min. It may be necessary to spike the reaction mix with additional enzymes. If the size of the labelled fragments is correct, add 5 μl of each stop solution and inactivate the enzymes by heating the vial(s) for 2–4 min at 94 °C.

5.2 Random primer labelling

Protocol 8

Random primer labelling

Equipment and reagents

 - 37 °C water-bath
 - Microcentrifuge
 - 95 °C heat block
 - 1 mM commercial fluor/hapten labelled dUTP
 - 10 ng–3 μg template DNA
 - Klenow fragment of DNA polymerase I
 - Random primers, six to nine bases long

 - 10 mM dNTP mixture: 10 mM each of dATP, dCTP, dGTP
 - 10 × Klenow buffer: 0.5 M Tris–HCl pH 7.5
 - 100 mM $MgCl_2$
 - 10 mM DTT
 - 0.5 mg/ml BSA
 - 0.5 M EDTA pH 8.0

Protocol 8 continued

Method

1 Make a dNTP mixture:
   - 192 μl ddH$_2$O
   - 2 μl of 100 mM dATP
   - 2 μl of 100 mM dGTP
   - 2 μl of 100 mM dCTP
   - 1.3 μl of 100 mM dTTP
   - 0.7 μl of 100 mM fluor/hapten dUTP

   Store the excess in aliquots at –20 °C.

2 Heat denature DNA template (0.1 μg/μl) by incubating at 95 °C for 5 min, then cool rapidly in an ice–water slurry.

3 Combine in order in a microcentrifuge tube on ice:
   - 8 μl of heat denatured DNA template
   - 2 μl of 10 × Klenow buffer
   - 2 μl of 10 × dNTP mixture
   - 2 μl random sequence primers
   - 5 μl ddH$_2$O
   - 1 μl (2 U) Klenow enzyme

4 Transfer to a 37 °C water-bath and incubate for 2–8 h (preferably 6 h).

5 Stop the reaction by adding 1 μl of 500 mM EDTA pH 8.0.

6 Check the size of the random primed products on a 1% agarose TAE gel.

5.3 RNA transcription labelling

RNA probes can be made using T4 RNA polymerase using the T3, T7, or SP6 transcription promoters. Frequently, linearized plasmid DNA containing the desired probe complement is used as the template. However, PCR fragments can be made using transcription promoter sequences at the 5′ end of the primers to obtain suitable transcription templates or promoter sequences can be ligated on to the end of the PCR products (Lig'nScribe, Ambion). RNA is synthesized by an *in vitro* transcription reaction in which UTP is replaced by a mixture of UTP/ amino-allyl UTP (usually at a 1:1 ratio). Transcription reactions are then phenol extracted, RNA is ethanol precipitated, resuspended in 1 × SSC, and unincorporated nucleotides are removed by two successive micro-spin columns. The RNA is then again ethanol precipitated and resuspended in water.

Protocol 9

RNA transcription labelling

Reagents

- Suitable plasmid with desired insert
- RNA polymerase: T3, T7, or SP6 RNA polymerase
- $10 \times$ T3, T7 buffer: 200 mM Tris–HCl pH 8.0, 40 mM MgCl$_2$, 10 mM spermidine, 250 mM NaCl
- Unlabelled nucleotide mix: 10 mM ATP, 10 mM CTP, 10 mM GTP, 7.5 mM TTP
- $10 \times$ SP6 transcription buffer: 200 mM Tris–HCl pH 7.5, 30 mM MgCl$_2$, 10 mM spermidine
- 1 mM Chromatide UTP (Molecular Probes) or equivalent
- RNase inhibitor (RNasin, Promega Corp.)
- 200 mM EDTA pH 8.0

Method

1 Linearize sufficient plasmid to have 500 ng template/reaction, cutting downstream of the insert and leaving a 5′ overhang. Clean up the template by two rounds of phenol/chloroform extraction, followed by ethanol precipitation.

2 Redissolve the plasmid in 10 mM Tris–HCl, 1 mM EDTA pH 8.0 to 500 ng/μl.

3 Dilute the RNA polymerase to 2 U/μl in $1 \times$ transcription buffer.

4 Dilute the RNasin to 1 U/μl.

5 Add together, mixing well prior to adding template and enzyme:
   - 1 μl unlabelled nucleotide mix
   - 1 μl Chromatide UTP
   - 2 μl of $10 \times$ transcription buffer
   - 1 μl RNasin (1 U)
   - 12 μl RNase-free dH$_2$O
   - 1 μl DNA template (500 ng)
   - 2 μl RNA polymerase (4 U)

6 Incubate at 37 °C (T3 or T7) or 40 °C (SP6) for 2 h.

7 Add 10 U of RNase-free DNase and continue incubation at 37 °C for 15 min.

8 Add:
   - 2 μl of 200 mM EDTA pH 8.0
   - 5 μl of 20 mg/ml yeast tRNA
   - 3 μl of 4 M LiCl
   - 75 μl cold 100% EtOH

9 Mix and freeze at –20 °C for at least 20 min.

10 Spin hard at 4 °C for 15 min.

11 Wash in 70% EtOH and repeat spin.

5.4 PCR labelling

Protocol 10

PCR labelling

Equipment and reagents

- Thermocycler
- 10 × PCR buffer: 500 mM KCl, 100 mM Tris–HCl pH 8.4, 15 mM $MgCl_2$
- 5 μM primer(s)
- Taq polymerase
- 2 mM dTTP

- 5 mM dNTP mixture: 5 mM each of dATP, dCTP, dGTP
- 1 mM commercial or custom (*Protocol 6*) fluor/hapten dUTP
- 1–10 ng template DNA

Method

1 Mix on ice in thin-walled PCR tubes:
   - 10 μl of 10 × PCR buffer
   - 6–8 μl of 50 mM $MgCl_2$ (if custom nucleotides are used)
   - 4 μl of 10 mg/ml BSA (if custom nucleotides are used)
   - 2–4 μl primer(s) (20–50 μM stock)
   - 4 μl of 5 mM dNTP mixture
   - Water to 100 μl

2 If using nucleotides prepared by *Protocol 6*, add:
   - 7 μl of 1 mM dTTP
   - 1.2–6.4 μl of 1 mM custom labelled dUTP
   - 2 μl DEAC (final 20 μM)
   - 3.5 μl R6G, TAMRA, TxR (final 30 μM)
   - 5.5 μl Cy3, Cy5 (final 50 μM)
   - 8 μl AMCA, FITC, BIO, DIG, DNP (final 60–70 μM)

3 If using commercial nucleotides, add:
   - 14 μl of 1 mM dTTP
   - 1.2–6.4 μl of 1 mM commercial labelled dUTP
   - 1.6 μl DEAC (final 20 μM)
   - 3 μl R6G, TAMRA, TxR (final 30 μM)
   - 5 μl Cy3, Cy5 (final 50 μM)
   - 6.5 μl AMCA, FITC, BIO, DIG, DNP (final 60–70 μM)

4 Add:
   - 1 μl Taq (5 U/μl stock)
   - 4 μl DNA template (1–100 ng)
   - Water to 100 μl

5 Cap tightly, place tubes in PCR machine, and run the following program of 30 cycles:

(a) 30–60 sec at 94 °C (denaturing).

(b) 50–60 °C for 30–60 sec (annealing).

(c) 68–72 °C for 0.5–4 min (extension).

6 Check the quantity of PCR product and the size on a 1% agarose TAE gel. If products are too long, treat with DNase and recheck size.

7 Probes made with custom nucleotides must be purified before use to remove the BSA as described in Section 6.2

6 Post-labelling DNA processing and purification

After labelling the DNA needs to be processed and purified before it is used in FISH (or some other applications).

6.1 DNase treatment (for FISH and other hybridization protocols)

To obtain labelled DNA fragments between 200–500 bp in size, a partial DNase treatment frequently is required. This can be accomplished by a short incubation at room temperature. The strength of DNase solutions may vary, so it is important that any batch of DNase is tested, by digesting the same DNA sample with the same amount of DNase for different periods of time (2–20 min). The sensitivity of DNA to DNase varies considerably depending on the DNA preparation and the amount of nucleotide analogue that it has incorporated (highly substituted DNA is more resistant to DNase digestion). The $10 \times$ DNase digestion solution is obtained by mixing 400 μl water, 4 μl of 1 M $MgCl_2$, and 1–2 μl of 3 mg/ml DNase stock solution (final DNase concentration in the reaction is about 1.5 μg/ml). The DNase solution can be added directly into the labelling reaction, followed by incubation for 10–15 min at room temperature, and is stopped by heating the sample for 2–3 min at 94 °C.

6.2 Removal of unincorporated nucleotides and BSA from the reaction mix prior to probe precipitation

For digoxigenin and rhodamine labelled probes: 1–2 μl proteinase K (2 mg/ml) is added to every 100 μl PCR labelling reaction and the vials incubated for 30 min at 37–45 °C. The reaction is stopped by adding the irreversible protease inhibitor Pefabloc SC (Roche Diagnostics), at a final concentration of 2–4 mM. The DNA can be simply precipitated and used. No DNA losses occur, and the DNA can be easily resuspended in hybridization buffer after precipitation.

For DNA labelled with all other fluors/haptens: labelled probe is subjected to phenol extraction, ethanol precipitation, and used.

7 Other labelling systems

7.1 Coupling of fluors or haptens to amine-modified nucleic acids

Many polymerases that will not incorporate fluor nucleotides will incorporate 5-(3-aminoallyl)-2′-deoxyuridine 5′-triphosphate. The resulting DNA strands can be labelled by chemically coupling fluors, a process which is much more efficient if the DNA is single-stranded. Kits are available that label by incorporation of amines during nick translation or reverse transcription reactions, followed by chemical coupling of fluor succinimidyl ester (ARES, Molecular Probes).

7.2 Other chemical coupling systems

New methods of labelling nucleic acids are constantly being developed. One example is the biotin-psoralen labelling system sold as kits by Ambion and Schleicher & Schuell.

7.3 Direct chemical coupling of fluors or haptens to proteins

The most common proteins labelled for use in nucleic acid detection are biotin-binding proteins (avidin) and immunoglobulins directed against haptens, fluors, or nucleotide analogues (BrdU). A wider selection of fluors is available for labelling proteins than nucleic acids, including Cy3.5, Cy5.5, Oregon Green, and the Alexa dyes. *Protocol 11* will work for the labelling of any protein. This protocol cannot be used for coupling of phycoerythrin to proteins. Phycoerythrin, which has excellent characteristics including a tremendous quantum yield, is typically coupled using heterobifunctional coupling reagents.

Protocol 11 describes the coupling of fluor to protein using fluor succinimidyl esters, which is the most satisfactory chemical form for protein coupling. Check for other ingredients in your protein. It is essential that the candidate protein is the only source of amines in the reaction mix. Other sources of primary amines, such as Tris buffer, must be removed from the protein solution by extensive dialysis prior to the coupling reaction. If possible try three different fluor/protein ratios, 1:7, 1:10, and 1:15, to determine the best ratio for your particular fluor/

Protocol 11

Succinimidyl ester labelling of proteins

Equipment and reagents

- G25 spin column (or other resin depending on the protein M_r)
- Fluor succinimidyl ester
- Dimethyl sulfoxide
- 0.1 M sodium bicarbonate pH 8.3
- 1.5 M hydroxylamine pH 8.5, or 2 M glycine pH 8.0 and 1 M Tris–HCl pH 8.0
- PBS
- 10 mg/ml BSA
- 20 mM sodium azide

Method

1 Dissolve the protein in 0.1 M sodium bicarbonate pH 8.3 at a concentration of 10 mg/ml.

2 Dissolve the fluor succinimidyl ester in DMSO at 10 mg/ml.

3 Add 0.1 vol. of fluor solution to the protein solution while stirring.

4 Incubate at room temperature for 1 h with continuous stirring.

5 Stop the reaction by adding 0.1 vol. of 1.5 M hydroxylamine pH 8.5.[a]

6 Add 0.1 vol. of 10 mg/ml BSA.

7 Separate the labelled protein from free fluor by applying to a spin column that is pre-equilibrated with PBS.

8 Add 0.1 vol. of 20 mM sodium azide and store the protein solution at 4 °C.

[a] Alternatively you can add 0.2 vol. of 2 M glycine pH 8.0 (20 mM final concentration) and 0.4 vol. of 1 M Tris–HCl pH 8.0 (20 mM final concentration).

protein combination. Alternatively, commercially available protein labelling kits that do three to five couplings can be used if only a few proteins are to be labelled with the same fluor.

References

1. Birnboim, G. C. and Doly, J. (1979). *Nucleic Acids Res.*, **7**, 1513.
2. Ioannou, P. A., Amemiya, C. T., Garnes, J., Kroisel, P. M., Shizuya, H., Chen, C., *et al.* (1994). *Nature Genet.*, **6** (1), 84.
3. Wang, M., Chen, X. N., Shouse, S., Manson, J., Wu, Q., Li, R., *et al.* (1994). *Genomics*, **24** (3), 527.
4. Albertsen, H. M., Abderrahim, H., Cann, H. M., Dausset, J., Le Paslier, D., and Cohen, D. (1990). *Proc. Natl. Acad. Sci. USA*, **87** (11), 4256.
5. Lengauer, C., Speicher, M. R., and Cremer, T. (1994). *Methods Mol. Biol.*, **33**, 85.
6. Guan, X. Y., Meltzer, P. S., and Trent, J. M. (1994). *Genomics*, **22** (1), 101.
7. Telenius, H., Carter, N. P., Bebb, C. E., Nordenskjold, M., Ponder, B. A., and Tunnacliffe, A. (1992). *Genomics*, **13** (3), 718.
8. Walker, G. T., Fraiser, M. S., Schram, J. L., Little, M. C., Nadeau, J. G., and Malinowski, D. P. (1992). *Nucleic Acids Res.*, **20** (7), 1691.
9. Compton, J. (1991). *Nature*, **350**, 91.
10. Guatelli, J. C., Whitfield, K. M., Kwoh, D. Y., Barringer, K. J., Richman, D. D., and Gingeras, T. R. (1990). *Proc. Natl. Acad. Sci. USA*, **87** (5), 1874.
11. Notomi, T., Okayama, H., Masubuchi, H., Yonekawa, T., Watanabe, K., Amino, N., *et al.* (2000). *Nucleic Acids Res.*, **28** (12), E63.
12. Matera, A. G. and Ward, D. C. (1992). *Hum. Mol. Genet.*, **1** (7), 535.
13. Bray-Ward, P., Menninger, J., Lieman, J., Desai, T., Mokady, N., Banks, A., *et al.* (1996). *Genomics*, **32**(1), 1.

14. Piatek, A. S., Tyagi, S., Pol, A. C., Telenti, A., Miller, L. P., Kramer, F. R., *et al.* (1998). *Nature Biotechnol.*, **16** (4), 359.

15. Kandimalla, E. R. and Agrawal, S. (2000). *Bioorg. Med. Chem.*, **8** (8), 1911.

16. Nielsen, P. E., Egholm, M., Berg, R. H., and Buchardt, O. (1991). *Science*, **254** (5037), 1497.

17. Henegariu, O., Bray-Ward, P., and Ward, D. C. (2000). *Nature Biotechnol.*, **18** (3), 345.

Chapter 3

Human chromosome mapping of single copy genes

Barbara G. Beatty

Department of Pathology, University of Vermont College of Medicine,
Room 206 Health Sciences Research Facility, Burlington, VT 05405, USA.

Stephen W. Scherer

Department of Genetics, Hospital for Sick Children, 555 University Avenue,
Toronto, Ontario M5G 1X8, Canada.

1 Introduction

One of the most common applications of FISH is the localization of single copy genes to a specific chromosomal band on metaphase chromosomes (1–3). Identifying the subchromosomal position of a gene in both normal and disease states can provide valuable information regarding its biological and/or clinical significance. In some situations, mapping a newly identified gene to regions of chromosomal deletion or amplification, to translocation breakpoints, or to a region near a gene known to be involved in a particular disease process (4–6) can supply essential clues which may lead to important clinical applications (see Chapter 9). Similarly, the use of FISH to map genes obtained from specific tissues or disease conditions by PCR-based cloning techniques (exon trapping, differential display) (7), to identify transgene insertion sites (8) or murine synteny (9), or to order two or more genes on a chromosome can provide information critical to understanding gene function.

Criteria for precise, reliable mapping results include clean high quality probes, well spread chromosomes with clear banding patterns, and optimized labelling and detection systems. Use of an appropriate epifluorescence microscope, digital camera, and corresponding software with attention to elimination of registration shifts during image capture all contribute to optimal signal-to-background ratios and precise mapping localization. In this chapter we present methods for identification and preparation of FISH probes and target DNA, and protocols for hybridization and detection systems suitable for mapping and ordering single copy genes.

2 DNA probes for FISH mapping

2.1 Identification of FISH probes from WWW sites

Over the past decade a wealth of information has emanated from the Human Genome Project providing a rich source of DNA sequences suitable for FISH mapping studies. The number of ESTs and genes that have been sequenced or partially sequenced has grown exponentially and databases containing this information are publicly available. Today, the best initial method for identifying to which chromosomal band a sequence of interest maps, is to use *in silico* methods. Confirmation may then be necessary using a genomic probe and FISH mapping methods. Bacterial cosmid, BAC, PAC, and YAC clones now exist for much of the genome but, to date, there is no one single WWW site showing the order and relationship of these clones with respect to each other. Moreover, the quality of these genomic clones for FISH probes is not always known. A general schema outlining how to identify clones for experimentation is presented below.

Sections (a) and (b) describe identifying clones based on cytogenetic position (most of these clones are also known to contain DNA marker(s) so they can be anchored to a physical or genetic map). Links to well-established FISH maps can be found in these sections containing information on pre-mapped clones. Based on cytogenetic localization additional clones can also be found.

In Sections (c) and (d) the methods of identifying clones specific for a given gene or DNA marker are described. The isolation of overlapping or adjacent clones to those described in Sections (a–d) might be required and the strategy to accomplish this can be found in Section (e). Finally, a list of WWW sites is given in Section (f) where the actual mapping reagents might be obtained. The availability, cost (some centres are government subsidized), and delivery time will vary from site to site so it is worth checking each one. Often the nomenclature describing the clones listed in the database will not be identical to that used in the respective reference library systems, so when ordering, it is important to provide as much information including clone name, source library, and accession number as possible.

(a) Pre-mapped FISH probes from chromosome-specific maps.
  i. http://www.gdb.org/hugo/: each chromosome has a link called 'FISH maps' under 'Chromosome Resources' which lists ordered and pre-tested genomic YAC, BAC, and PAC probes.

(b) Additional clones based on cytogenetic localization.
  i. Look at WWW site listed in Section (a).
  ii. Identify genomic clones with known cytogenetic localizations from the Genome Database (GDB) at http://www.gdb.org/gdb/regionSearch.html.
  iii. If the desired genomic clone is not obtained using the same GDB site or GeneCards (http://bioinfo.weizmann.ac.il/cards/index.html) identify DNA markers or genes based on their cytogenetic location and analyse them as described in Section (c) or (d).

iv. the Genetic Location Database (http://cedar.genetics.soton.ac.uk/
public_html/gmap.html) is another source to identify genes and DNA
markers by cytogenetic position to be analysed as in Section (c) or (d).

(c) Text search to retrieve a genomic clone specific for a gene or DNA marker

   i. BACs, PACs, or cosmids.

Genome Database:	Locus Link:
(http://www.gdb.org)	(http://www.ncbi.nlm.nih.
–enter gene or marker name	gov/LocusLink/)
click gene or amplimer to get	–enter gene name
to 'Nucleic Acid Links'	–click on 'Entrez Map Viewer'
–link to DNA sequence database	–link to DNA sequence database
(e.g. Genbank sample, NCBI)	–scan record to identify clone name
–scan record to identify clone name	

   ii. YACs.

Whitehead/MIT (http://carbon.wi.mit.edu:8000/cgi-bin/contig/phys_map)
–enter using either DNA marker name or YAC clone name
–scan record to identify clone as well as contig to obtain overlapping
clones along the chromosome

(d) DNA sequence search for corresponding genomic clone.

   i. Retrieve desired sequence of gene or DNA marker by text search of:
(http://www.ncbi.nlm.nih.gov:80/entrez/query.fcgi?db=Nucleotide).

   ii. Copy the desired accession number or DNA sequence and paste into:
http://www.ncbi.nlm.nih.gov/blast/blast.cgi.

   iii. Screen the: Database NR (default) to find 'Finished' sequences.
                  Database HTGS for 'rough draft' sequence.
                  Database GSS to find overlapping BAC ends.

   iv. Scan records to identify those containing corresponding genomic clones.

(e) Identifying overlapping and/or nearby clones.

   i. To identify overlapping BACs based on a known BAC clone name or
DNA sequence see: http://www.TIGR.org/tdb/humgen/bac_end_search/
bac_end_intro.html.

   ii. To identify overlapping BACs based on fingerprint patterns see: http:
//genome.wustl.edu/gsc/human/human_database.shtml.

   iii. Ensembl (http://www.ensembl.org/), NCBI (http://www.ncbi.nlm.nih.gov/
cgi-bin/Entrez/hum_srch?chr=hum_chr.inf), and UCSF (http://genome.
ucsc.edu/) are also generating overlapping BAC maps based on DNA
sequence alignment. These three sites also offer good entry points to get
clones based on cytogenetic localizations.

   iv. To identify overlapping YACs see Section (c).

(f) Obtaining clone(s) for experiments.

   i. Canada: The Centre for Applied Genomics (http://tcag.bioinfo.sickkids.
on.ca/)—BACs, PACs, cosmids, YACs.

   ii. Germany: DHGP (http://www.rzpd.de/)—cosmids, BACs, PACs, YACs.

   iii. Italy: TIGEM (http://www.spr.it/iger/home.html)—BACs, PACs, YACs.

   iv. UK: HGMP Resource Centre (http://www.hgmp.mrc.ac.uk/)—cosmids, BACs, PACs, YACs.

   v. UK: Sanger Center (http://www.sanger.ac.uk/HGP/)—BACs, PACs.

   vi. USA: Research Genetics (http://www.resgen.com/)—BACs, PACs, YACs.

   vii. USA: Genome Systems (http://reagents.incyte.com/)—BACs, PACs, YACs.

   viii. USA: Pieter de Jonge's lab (http://www.chori.org/bacpac/)—BACs, PACs.

2.2 Preparation of probes for FISH mapping

FISH mapping has been used to localize single copy DNA sequences, groups of DNA sequences, gene families, pseudogenes, translocation breakpoints, and regions of gene loss or amplification to specific subchromosomal banding regions. In all cases, the probe type, size, and purity are each critical features that must be taken into consideration in order to obtain precise mapping results. The two most common types of probes used for mapping studies are:

(a) The single copy genomic DNA probes: plasmids, cosmids, PACs, BACS, and YACs. Single copy genomic probes range in size from 5 kb–1 Mb and their signal strength is usually dependent on probe size. These probes contain regions of common repeat sequences (SINE, LINE, Alu) which must be pre-annealed with Cot-1 DNA prior to hybridization (10) to minimize high non-specific background signal across the genome. Genomic probes can be labelled directly or indirectly and the method of choice usually depends on the size of the probe (see Chapter 2).

(b) The unique sequence DNA probes: cDNAs, and expressed sequence tags (ESTs). Unique sequence probes (<5 kb) do not usually require suppression by Cot-1 DNA but their signal strength can often be unpredictable since it is related to the size and structure of the gene target. Transcribed sequences may comprise a number of small exons or may have larger runs of contiguous exonic transcript. Because of the complexity and uncertainty of ESTs and cDNAs, it is best to assess such probes empirically. These probes usually require indirect labelling and detection amplification (see Section 7, *Protocol 10*) for visualization.

For both types of DNA probes, purity (no RNA) and size of labelled fragments (100–400 bp) (see Chapter 2) are critical in producing the best signal-to-background ratios. We have found CsCl or Qiagen MiniPrep column purification to be the most successful in producing probes suitable for mapping. Once labelled (see Chapter 2), the probes are ethanol precipitated in the presence of excess non-human carrier DNA such as salmon sperm or herring testes DNA to reduce non-specific binding to both target DNA and the slide during hybridization. Carrier DNA must have the same fragment size distribution as shown by the probe, and can be sonicated in the laboratory or purchased commercially (Life Technologies) with a size range of 500–2000 bp. For single copy genomic probes, a second ethanol precipitation in the presence of excess unlabelled human Cot-1 is also required.

Protocol 1

Ethanol precipitation of labelled probe DNA

Equipment and reagents

- Refrigerated microcentrifuge (Eppendorf)
- 2 μg labelled probe
- 3 M sodium acetate
- 70% and 100% ethanol (cold)
- Salmon sperm DNA (5 μg/μl) (Sigma)
- Cot-1 human DNA (1 μg/μl) (Life Technologies)
- Hybridization buffer: 50% high grade deionized formamide (Fluka Chemical Corp.), 2 × SSC, 10% dextran sulfate; or Hybridization buffer (Dako Corp)

A. Precipitation with salmon sperm DNA

1 To a 1.5 ml Eppendorf tube containing the labelled probe, add salmon sperm DNA (50 μg for each μg labelled DNA).[a]

2 Add 0.1 vol. of 3 M Na acetate and mix well (final concentration 0.3 M).

3 Add 2 vol. of cold 100% ethanol, mix by inverting the tube, and allow the tube to sit for at least 20 min at –70 °C (or overnight at –20 °C).

4 Centrifuge the DNA at 2000 g for 20 min at 4 °C.

5 Carefully decant (or gently aspirate) the supernatant into a tissue and add 0.5 ml of cold 70% ethanol. Invert the tube once gently.

6 Repeat step 4, carefully remove the supernatant as in step 5, and allow the pellet to air dry for 1–2 h or vacuum dry for 15 min.[b]

7 Add autoclaved distilled water to the dry pellet to give a final concentration of 10 ng/μl and vortex or flick tube to dissolve the DNA completely.[c]

B. Precipitation with Cot-1[d]

1 For genomic probes, add Cot-1 DNA[e] to the amount of probe required for mapping,[f] and adjust the final volume to 100 μl with distilled water.

2 Repeat part A, steps 2–6.

3 Store at –20 °C as a dry pellet or in hybridization buffer (10–15 μl per slide).

[a] Check the amount of labelled probe DNA prior to precipitation by running ssDNA standards of known concentration together with the probe in the TBE 1% agarose gel used to check probe fragment size.

[b] Cover the tube with tissue paper to prevent the pellet from being sucked out.

[c] The DNA can be stored at –20 °C until the second precipitation.

[d] If hybridizing two probes simultaneously, combine the probes and add the amount of Cot-1 required for each individual probe. Precipitate the probes together and resuspend in hybridization buffer (10–15 μl per slide).

[e] 3–5 μg Cot-1/100–300 ng probe (i.e. per slide) is sufficient for most probes. YAC probes require more, ~25 μg. The amount of Cot-1 required will vary depending on the background and signal strength. Too much Cot-1 will result in low chromosomal background but a weaker signal; too little Cot-1 will results in high chromosomal background but a stronger signal.

[f] We suggest optimizing each new probe using a range of probe concentrations (genomic DNA: 100–300 ng/slide; cDNA: 300–600 ng/slide) and duplicate slides.

3 Target DNA preparation

3.1 Metaphase chromosomes

Preparation of high quality target DNA is also essential for accurate FISH mapping. Sub-band localization of single copy probes (genes) (see *Figure 2*) is best performed on metaphase chromosomes with well defined banding patterns. For most mapping studies, suitable metaphase chromosomes can be prepared from short-term cell cultures using slight modifications of conventional cytogenetic techniques (11). Mutagen-stimulated lymphocytes from human peripheral blood are the most common source of human metaphases, however, lymphoblasts, fibroblasts, tumour lines, or amniocyte cultures, can also be used. Although not necessary for most mapping work, cell synchronization (*Protocol 3*) may be used to increase the number of cells in metaphase and is required if using chromosome R-banding (12). In this section we describe protocols for culturing and harvesting peripheral blood lymphocyte cultures for production of high quality metaphase chromosome targets.

3.1.1 Preparation of metaphase chromosomes from cell culture

Preparation of chromosomes from short-term human peripheral blood cultures entails four main steps:

(a) Mitogenic stimulation of the peripheral T lymphocytes by phytohaemagglutinin (PHA).

(b) Arrest in metaphase using a mitotic inhibitor (colcemid).

(c) Hypotonic treatment of the cells to reduce cytoplasm and facilitate chromosome spreading.

(d) Fixation to arrest cell function, maintain good cell morphology, and prevent further swelling.

Human peripheral blood lymphocytes are obtained from heparinized whole blood and should be cultured within three to four days. Addition of PHA to the culture medium stimulates T lymphocytes to divide, with a maximum mitotic activity usually being reached between 60–70 h. Cell culture times therefore usually range from 48–72 h. The cells are then arrested in metaphase (when DNA is structurally condensed into distinct chromosomes) by addition of the spindle inhibitor colcemid. In metaphase, the nuclear membrane disintegrates and disruption of the spindle fibres allows the chromosomes to break free of the metaphase plate and disperse within the cell. Both the amount and the exposure time of colcemid are critical factors in producing good metaphase chromosomes.

Optimal subculture dilutions should be determined individually for other sources of cells. A cell density of about 1×10^6 cells/ml of medium is usually appropriate for harvesting. Colcemid is added 30–45 min prior to harvest. The mitotic index of adherent cells (fibroblasts) can be increased by subculturing 48 h and feeding 24 h prior to harvest. Adherent cells should be harvested at 50–80% confluency and colcemid added to the media 4–6 h (or overnight) prior to harvesting. The concentration of colcemid and length of exposure time will depend on the particular cell line (see *Protocol 2*, footnote c).

Protocol 2

Culture and harvest of peripheral blood lymphocytes

Equipment and reagents

- CO_2 incubator
- Low speed centrifuge
- Vacutainer blood collection tubes containing sodium heparin
- T25 tissue culture flasks
- Cell culture medium[a] (100 ml): 80 ml RPMI 1640 medium containing Hepes (Life Technologies), 20 ml fetal calf serum, 1 ml
- of 2 mM L-glutamine (Invitrogen or Sigma), 1 ml of 100 U/ml penicillin/ 100 mg/ml streptomycin (Invitrogen or Sigma), and 1 ml PHA (M-form, reconstituted in 10 ml sterile distilled water just prior to use, Invitrogen)
- 10 μg/ml colcemid[b] stock solution (Invitrogen or Sigma)

Method

1   Collect 10 ml venous blood into a Vacutainer tube containing sodium heparin.

2   Add 10 ml whole blood to 100 ml cell culture medium and aliquot 10 ml per T25 tissue culture flask.

3   Incubate the flasks in a 37 °C CO_2 incubator for 60–72 h.

4   Approx. 24 h prior to harvesting the cells, carefully remove 5 ml of medium from the flasks. Be careful not to resuspend any cells. Replace with 5 ml of fresh medium.

5   Add 100 μl sterile colcemid solution to each flask (10 ml) for a final concentration of 0.1 μg/ml, and continue incubation for 20–30 min at 37 °C.[c]

6   Harvest the cells by transferring them to a 15 ml conical centrifuge tube and spin at 200 g for 5 min. Continue with *Protocol 5*.

[a] Cell culture medium can be stored at 4 °C for one month or for much longer periods at –20 °C.

[b] Colcemid should be stored at –20 °C in 50 μl aliquots and thawed just prior to use. Colcemid is toxic and contact with skin should be avoided.

[c] Timing is critical as too long a treatment with colcemid will result in short, condensed chromosomes and tight metaphase spreads.

3.1.2 Synchronized cell cultures

The number of cells in metaphase can be increased by synchronization of cell cultures. One way to achieve this is by addition of methotrexate (amethopterin) which inhibits thymidine incorporation into DNA, resulting in early S phase growth arrest. On removal of methotrexate and addition of thymidine to the medium, the cells resume DNA synthesis and continue through the cell cycle in a synchronized fashion. Replacing thymidine with the thymidine analogue 5-bromodeoxyuridine (BrdU) following cell arrest, results in BrdU incorporation into late S phase replicating DNA, producing longer chromosomes and an R-banding pattern when the chromosomes are stained with propidium iodide (PI) (12, 13). Since the timing of colcemid addition is critical, some knowledge of the cell cycle timing in the presence of PHA is required in order to catch the cells as they enter metaphase.

Protocol 3

Cell synchronization and BrdU incorporation

Equipment and reagents

- Cell culture medium (see *Protocol 2*)
- 10 mM methotrexate (Sigma)
- Cell culture release medium: cell culture medium (*Protocol 2*) containing 10 μM thymidine or 100 μM BrdU[a]

Method

1 Culture cells as in *Protocol 2*, steps 1–3.

2 Resuspend cells by swirling and add 100 μl of 10 μM methotrexate per 10 ml cell culture (0.1 μM final concentration).

3 Return cells to incubator for 15–18 h (overnight).

4 Transfer cells from each flask to a 15 ml centrifuge tube and spin at 200 g for 8 min at 37 °C.

5 Remove all but 0.5 ml supernatant and resuspend the pellet by flicking the tube.

6 Add 10 ml culture medium (warmed to 37 °C) to each tube. Mix well by inversion.

7 Centrifuge as in step 4.

8 Remove supernatant, resuspend each pellet in 10 ml of cell culture release medium, and transfer to T25 flasks. Return flasks to the incubator for 3.5–7 h with the lid loosely capped.

9 Repeat *Protocol 2*, steps 5 and 6, using a shorter colcemid incubation time of 10–12 min.[b]

[a] Deoxycytidine (100 μM) can be added to decrease the toxic effects of BrdU. After addition of BrdU cover the tubes (with foil) as BrdU-substituted chromosomes are susceptible to breakage when exposed to certain wavelengths of light (14).

[b] For longer chromosomes, 10 μl of a 10 mg/ml ethidium bromide solution can be added to each flask 2 h prior to harvest. Note: ethidium bromide treatment may have a detrimental effect on chromosome morphology.

3.2 Mapping with interphase nuclei

Interphase DNA mapping is best performed on G0 phase diploid fibroblast nuclei. This approach can be used for distance determination (two or more probes), or ordering (three or more probes), of probes that are not well separated on metaphase chromosomes (i.e. are <1–2 Mb apart). The best source of inter-phase nuclei is confluent fibroblast cultures arrested in G0–G1 phase. Uncultured peripheral blood lymphocytes are also a source of interphase nuclei, but these preparations often contain G2 nuclei. It is best to avoid nuclei in G2 phase (tetraploid) as signal interpretation becomes more difficult. For general informa-tion on culturing fibroblasts see *The AGT cytogenetics laboratory manual*, Chapter 4 (15). Fibroblast nuclei have the advantage of being large and flatten well follow-ing hypotonic treatment with the result that the FISH signals tend to all be in the same plane making visualization and interpretation easier. For more detailed information on using interphase nuclei for FISH mapping, see *Genome analysis: a practical approach*, Chapter 9 (16).

Protocol 4

Preparation of fibroblast G0–G1 interphase nuclei

Reagents

- Fibroblast culture[a]
- Citrate saline: 0.44% (w/v) Na citrate, 1% (w/v) KCl, warmed to 37 °C
- Trypsin, EDTA (Life Technologies): 5% in citrate saline warmed to 37 °C
- 0.075 M KCl warmed to 37 °C

Method

1 Grow fibroblast cells in T25 flasks under standard culture conditions until they reach confluency.

2 Allow the cells to sit at 37 °C for about seven days or until no mitotic cells are visible.

3 Pour off growth medium into 15 ml conical centrifuge tube.

4 Add 5 ml pre-warmed citrate saline, incubate for 5 min, and pour into the same centrifuge tube.

5 Add 3 ml trypsin solution and incubate until remaining adherent cells are resus-pended (1–2 min). Add to same centrifuge tube.[b]

6 Pellet the cells by centrifugation at 200 g for 5 min at room temperature.

7 Remove supernatant, flick the tube to partially resuspend the pelleted cells, and add 12 ml of 0.075 M KCl (warmed to 37 °C).

8 Mix by inverting the tube several times and incubate for 10 min at 37 °C.

9 Continue as in *Protocol 5A*, step 6 onward.

[a] Fibroblast cultures can be obtained from a skin biopsy or from ATCC. Male cells are preferable for mapping.

[b] The medium in the tube will neutralize trypsin activity.

3.3 Hypotonic treatment and fixation

Harvested cells are next exposed to a hypotonic solution of KCl or to diluted cell culture medium to cause water uptake which swells the lymphocytes making them more fragile. Hypotonic swelling facilitates dispersal of the chromosomes and lyses any remaining red blood cells. Addition of fixative causes nucleoprotein crosslinks to form, and also dehydrates cells and preserves them in their swollen state. Fixative also enhances the attachment of chromosomes and nuclei to the slides and improves DNA accessibility. The cells can be pelleted and stored in fixative at –20 °C for several months provided they are capped tightly. Frequent usage and exposure to room temperature air results in methanol evaporation and a relative increase in acid concentration as well as uptake of moisture causing esterification and contamination of the fixative. This can be minimized by resuspending the pellet in fresh fixative each time and aliquoting fresh suspension into 1 ml nylon cryotubes (Nunc) for storage (–20 °C).

Protocol 5

Hypotonic treatment and cell fixation

Equipment and reagents

- 37 °C water-bath or incubator
- Hypotonic solution:[a] 0.075 M KCl, or a 1:1.4 medium/water solution consisting of 5 ml culture medium (including 10% fetal calf serum) and 7 ml sterile distilled H_2O
- Fixative: 3:1 methanol/glacial acetic acid[b]
- Phosphate-buffered saline (PBS) at ambient temperature

A. KCl hypotonic treatment

1 Following centrifugation (*Protocol 2*, step 6), carefully aspirate all but 0.2–0.5 ml of the culture medium from the cell pellet and resuspend the pellet by flicking the tube several times to prevent clumping.

2 Add 10 ml PBS and centrifuge at 200 g for 5 min at room temperature.

3 Repeat step 1.

4 Slowly add 10 ml warm (37 °C) 0.075 M KCl solution: add the first 1 ml dropwise.[c] Mix after each addition by flicking the tube or by gentle vortexing.

5 Incubate in a 37 °C water-bath for 10 min.

6 Add five drops of fixative, then gently invert the tube two or three times to mix, and centrifuge the cells as in step 2.[d]

7 Repeat step 1 being very careful not to disturb the cell pellet.

8 Slowly add 10 ml of fixative, adding the first 1 ml dropwise (taking about 2 min),[e] and mixing gently after each addition.

9 Allow the suspension to sit at room temperature for 10 min. Centrifuge at 200 g for 5 min.

10 Remove the supernatant by carefully decanting or aspirating and add 10 ml of fresh fixative. This second fixation can be performed rapidly. Centrifuge as in step 9.

11 Repeat step 10 twice and store at –20°C.

B. Medium/water hypotonic treatment

1 Resuspend the cell pellet in 5 ml medium and add 7 ml sterile distilled water.

2 Mix by gently inverting the tube two or three times, and incubate in a 37°C water-bath for 10 min.

3 Continue as in part A, step 6.

[a] Both solutions should be pre-warmed to 37°C before use.

[b] The fixative should be prepared fresh each time, preferably from an unopened bottle of methanol to minimize atmospheric water absorption.

[c] The lymphocytes become very fragile during the hypotonic shock. Initial slow addition of the KCl helps prevent them from lysing.

[d] After centrifugation of this first fixation you should notice that the pellet has become more transparent and has doubled in size. This is a sign of adequate cell swelling following the hypotonic treatment. The suspension may turn blackish as the fixative lyses any remaining red blood cells. This is removed by subsequent washings with fixative.

[e] Initial slow addition of the fixative is critical as the hypotonic cells are very fragile.

4 Slide preparation

Slide making is more of an art than a science and each laboratory tends to develop its own approach. It is thought that swollen cells suspended in fixative immediately collapse when dropped onto a clean glass slide, due to loss of surrounding liquid. As the fixative evaporates, the cells flatten, the chromosomes spread out, and the thin cell membrane becomes invisible. Humidity, temperature, and overall drying time are critical features affecting chromosome spreading—with faster drying times resulting in less optimal spreading. Cells stored in fixative should be rinsed in fresh fixative just before use as mentioned above, as pH and water content may have changed during storage.

Chromosome spreads suitable for high efficiency hybridization should have minimal cytoplasm. Residual cytoplasm reduces FISH signals and results in high background. The chromosomes should be dark grey with sharp borders when viewed under phase-contrast. Dark black refractile, hollow, or pale grey chromosomes do not hybridize well. Chromosomes that are well spread out (not over-lapping) and demonstrate good banding resolution (not too condensed or fluffy) are appropriate for accurate mapping (see *Figure 1*). In recent years, controlled environmental evaporation chambers (Thermotron) designed for more repro-ducible slide making have gained in popularity in clinical laboratories (see Chapter 9). Alternatively, an inexpensive domestic humidifier in a hood can be used to increase humidity when the local climatic conditions are too dry.

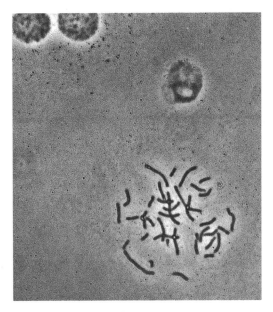

Figure 1 Metaphase chromosomes suitable for FISH. Human peripheral blood lymphocytes were cultured for 72 h, arrested in metaphase by addition of colcemid (1 μg/ml), and harvested as described in *Protocol 2*. The hypotonic treatment and fixation was performed according to *Protocol 5*, and slides were prepared according to *Protocol 6*. Under phase-contrast, the chromosomes are dark grey, well defined, and show few overlaps. The cytoplasm is barely visible.

Protocol 6

Slide preparation

Equipment and reagents

- Glass microscope slides
- Pasteur pipettes, ice bucket
- 1:1 concentrated HCl/95% ethanol solution

Method

1   Clean slides by immersing them in a 1:1 HCl/ethanol solution for 5 min. Place under running tap-water.

2   Store in distilled water at 4 °C, change to fresh distilled water (4 °C) just before use.

3   Resuspend the fixed cells (*Protocol 5*), in sufficient fresh fixative to give a faintly turbid or white suspension.

4   Remove slide from cold distilled water, and using a glass Pasteur pipette or Pipetman™ drop one drop of cell suspension from a height of about 1–3 cm on to the slide and tilt at an angle of about 45°.

Protocol 6 continued

5   Allow the drop to spread out and run down the slide. Let the slide dry by resting it vertically at a 45° angle.[a,b]

6   Check the slides under a phase-contrast microscope (×10 objective) to evaluate mitotic index, metaphase quality (spreading), and chromosome morphology.[c]

7   Age the slides at least two days at room temperature and preferably one to two weeks before use with single copy probes.

8   Store the slides at room temperature for two to four weeks or at −20 °C in a sealed container with desiccant for several months.

[a] Alternatively, wipe excess fluid from slide, breathe onto slide to facilitate chromosome spreading, and place slide on a hot plate (~55–60 °C, or just hot enough to touch) for 5–10 sec to dry. Remove any remaining drops of suspension by tapping onto a gauze pad.

[b] Humidity is critical. Too low a humidity impairs the quality of metaphase chromosomes. Too high a humidity and slow drying time results in 'ghost' chromosomes which appear empty with a dark outline.

[c] Metaphases should be two or three per field (×10), spread out, and have no visible cytoplasm.

4.1 Target slide pre-treatment

Slides can be pre-treated to remove cytoplasmic RNA or protein to reduce non-specific binding to nucleic acids and protein present on the slide. RNase pre-treatment is often required to reduce background RNA staining when using counterstains such as propidium iodide (PI) (R-banding) which stain both DNA and RNA, but is not usually necessary with DAPI counterstain. Proteinase K removes cytoplasmic and chromosomal proteins that may reduce target access and compromise visualization of small probes such as plasmids or cDNAs. Following pre-treatment, the slides may be put through an ethanol dehydration series to improve chromosome morphology and hybridization efficiency.

Protocol 7

Target slide pre-treatment

Equipment and reagents

- 37 °C incubator
- 2 × SSC: dilute from 20 × SSC stock (3.0 M NaCl, 0.3 M Na citrate pH 7.6)
- DNase-free RNase A (Roche): 100 μg/ml in 2 × SSC
- Proteinase K (Sigma): 0.06 mg/ml in 2 mM $CaCl_2$, 20 mM Tris pH 7.4
- Cold ethanol series: 70%, 90%, 100% (on ice)

A. RNase pre-treatment

1   Allow slides stored at −20 °C to reach room temperature before opening the storage container.[a]

2 Add 100 µl of RNase to a 24 × 40 mm coverslip, place slide carefully on the cover-slip, and turn over. Incubate in a humidified box (line with blotting paper soaked with distilled water) at 37 °C for 0.5–2 h (not critical).

3 Remove the coverslip gently and wash in three changes of 2 × SSC for 3 min each at room temperature.

4 Proceed to part B, or dehydrate the slides in the cold 70%, 90%, and 100% ethanol series for 3–5 min each. Drain the slides but don't let them dry out between trans-fers. Following the final ethanol dehydration, air dry the slides in a vertical position.

B. Proteinase K pre-treatment[b]

1 Incubate the slides in a Coplin jar containing the proteinase K solution (pre-warmed to 37 °C) at 37 °C for 7 min.

2 Wash as in part A, step 3 and the dehydration part of step 4.

[a] This avoids condensation from forming on the slides and destroying the metaphases.
[b] This can also be performed following denaturation (*Protocol 8*).

5 Denaturation and hybridization of probe and target DNA

Heating double-stranded DNA in the presence of formamide destabilizes the DNA and permits strand dissociation to occur at a lower temperature thus main-taining chromosome morphology. The formamide used for denaturation and hybridization must be of high quality, be deionized, and have a pH of 7.0. Oxida-tion of deionized formamide causes DNA depurination, thus proper preparation and storage (–20 °C) is critical.

Single copy genomic probes (plasmids, cosmids, PACs, BACs, or YACs) requir-ing pre-annealing with Cot-1 DNA, should be denatured prior to the target DNA. The target slides can then be denatured and dehydrated during the pre-annealing process. Unique sequence probes (cDNAs) or repeat sequence probes (alphoid, satellite III repeats) not requiring pre-annealing, should be denatured after the slides. The latter probes should be placed on ice immediately after denaturation to prevent any reannealing, and added to the target slides as quickly as possible.

Target slide DNA denaturation time is critical and may vary with the age of the slides. Over-denatured chromosomes appear 'fluffy' and demonstrate poor quality banding, while under-denaturation reduces hybridization efficiency. Denatured target slides can remain at room temperature following dehydration while the probes are being denatured.

Most probes are hybridized overnight at 37 °C, however, some larger genomic probes, alpha satellite, and repetitive probes often produce sufficient signal after 1–3 h of hybridization. In addition there can be marked differences in the stringency requirements for hybridization (discussed below) with different types of probes. Maintaining a humid environment during hybridization prevents the slides from drying out and compromising hybridization efficiency.

Protocol 8

DNA denaturation and hybridization

Equipment and reagents

- 37 °C oven
- 75 °C or 95 °C water-bath
- Light-resistant plastic box lined with gauze or blotting paper soaked in distilled water
- 2 × SSC (see *Protocol 7*)
- Denaturing solution: 70% formamide (Fluka Chemical Corp.) in 2 × SSC

- Hybridization buffer: 50% (v/v) deionized formamide[a] (Fluka Chemical Corp.), 2 × SSC, 10% (w/v) dextran sulfate[b] (Sigma); or Hybrisol VI (65% formamide) or VII (50% formamide) (Ventana)
- 70%, 90%, and 100% ethanol series (at –20 °C)
- Rubber cement

A. Probe denaturation

1 Resuspend probe pellet in hybridization buffer (pre-warmed to 37 °C) to a concentration of 5–20 ng/μl.[c] Mix well and leave at room temperature until ready to denature.

2 For single copy genomic probes, place in a 75 °C water-bath for 5 min. Remove and incubate (pre-annealing step) for ~1 h at 37 °C. YACs may require a longer pre-annealing time of 1–2 h.

3 For unique sequence and most repeat sequence probes, incubate the probe for 10 min in a 95 °C water-bath then place directly on ice.

B. Target DNA denaturation

1 Heat freshly prepared denaturation solution in a glass Coplin jar placed in a water-bath heated to 70 °C +1 °C for each slide.[d]

2 Place the dehydrated slides in the denaturing solution for 2 min (this step is critical). Remove slides and immediately put into a Coplin jar containing cold 70% ethanol that has been kept at –20 °C and repeat the dehydration series in a cold ethanol series (*Protocol 7A*, step 4). Air dry the slides in a vertical position following the final ethanol dehydration.

C. Hybridization

1 Pre-warm slides in a 37 °C oven 5–10 min prior to use (optional).

2 Remove the probe from the oven (if pre-annealed) or ice bath (if not pre-annealed), mix, and spin briefly in a picofuge.

3 Add 12–15 μl of probe to coverslip, cover with a slide, and turn over. Gently press out all air bubbles and seal with rubber cement.

4 Place the slides in a sealed moist container. Place container in a ziploc™ plastic bag (optional).

Protocol 8 continued

5 Incubate overnight in a 37 °C oven.[e]

[a] High grade deionized formamide (Fluka, Sigma) can be purchased but is expensive. Lower grade formamide can be used but must be deionized prior to use with a mixed-bed ion exchange resin (Bio-Rad AG501-X8, 20–50 mesh), and filtered twice through Whatman No. 1 filter paper. Deionized formamide can be aliquoted and stored at –20 °C for up to one year. Formamide should be thawed just prior to use. Formamide is a teratogen and thus should be handled with caution and a fume hood should be used. For repeat sequence probes (α satellites) increase the concentration of formamide to 65% to minimize cross-hybridization.

[b] A stock 50% (w/v) dextran sulfate solution, 25 g dextran sulfate (Sigma) in 50 ml distilled water, can be prepared, aliquoted, and stored at –20 °C indefinitely.

[c] For smaller probes, higher concentrations may be required.

[d] To compensate for the drop in temperature on adding the slides, the temperature of the bath (and formamide solution) is increased 1 °C per slide. Test the temperature of the denaturing solution by inserting a clean thermometer directly into the solution in the Coplin jar. The pH of the formamide solution should remain at 7.0. Long incubation times at 70 °C or more may change pH, making the solution less effective. Slides should be used the same day they are denatured.

[e] Maintaining hybridization temperature at 37 °C is critical. Lower temperatures may result in increased cross-hybridization (especially with repeat sequence probes) and higher temperatures may decrease signal intensity.

6 Post-hybridization washes

Following hybridization, unbound DNA is removed by washing the slides in a slightly higher stringency solution than that used for hybridization in order to remove weakly bound probe. Washing can be done using a low salt solution at high temperature or a formamide solution at a lower temperature for a longer time. Since both hybridization and washing stringency can affect the signal specificity and intensity, it is important to determine the appropriate hybridization and wash conditions for each probe type in order to obtain optimal signal-to-background ratios. Stringency can be increased by either lowering the salt concentration, or by increasing the temperature, the concentration of formamide, or the duration time of the washes. Too high a stringency may decrease signal, whereas too low a stringency may increase non-specific hybridization background.

Protocol 9
Post hybridization washes

Equipment and reagents
- Water-bath at 42 °C
- Wash A (formamide wash): 50% formamide in 2 × SSC pH 7.0
- Wash B: 2 × SSC pH 7.0 (see *Protocol 7*)

Protocol 9 continued

Method

1 Allow three Coplin jars containing wash A and three Coplin jars containing wash B to reach 42 °C in a water-bath before removing the slides from the incubator.

2 Remove the rubber cement from around the coverslip using tweezers and allow the coverslip to come off in wash A.[a]

3 Wash the slides in wash A for 3 × 5 min, followed by 3 × 5 min washes in wash B.[b] Shake gently at each stage and drain the slides briefly between washes.[c]

[a] Do not allow the slides to dry out at any time during the wash or detection procedures.

[b] If the background is too high, or the signal is too weak, increase or decrease the stringency respectively. Some probes, such as alpha satellite probes, may need a higher wash A stringency (1 × SSC) and a higher wash B temperature (60 °C) if there is significant cross-hybridization.

[c] Slides may also be washed at 72 °C for 5 min in an appropriate concentration of SSC: 1 × SSC for single copy probes and chromosome paints, and 2 × SSC for unique sequence probes.

7 Immunodetection

The method used for probe visualization depends on the size and nature of both the probe and the target, and on the method chosen for labelling the probe. Larger genomic probes such as PACs, BACs, or YACs directly labelled with a fluorochrome-tagged dNTP (17, 18) can usually be visualized immediately following the post-hybridization washes. However, smaller single copy probes are more often labelled with a hapten such as biotin or digoxigenin (DIG) and detected by one or more layers of a fluorochrome-conjugated reporter molecule as outlined in *Protocol 10*.

A higher degree of signal amplification utilizing biotinylated tyramime (19, 20) has recently been used with some success to map small PCR labelled probes <1 kb in size (21). In this method, horseradish peroxidase bound to an antibody (anti-DIG) or hapten (avidin), catalyses the deposition of biotinylated tyramine at the site of hybridization of DIG labelled or biotinylated probes respectively. The biotinylated tyramine is then detected with fluorochrome labelled avidin. This technique is now patented by DuPont and manufacturers' protocols are available in TSA™ detection kits from NEN™ Life Science Products (http://www.nenlifesci.com) and Dako Corporation (http://www.dakousa.com). For a review of tyramide amplification and its use in *in situ* hybridization see Speel *et al.* (22). The choice of fluorochrome(s) for probe detection will depend on the filters available and must be compatible with the counterstain (see Chapter 2).

Protocol 10

Detection of probes labelled with biotin or DIG

Equipment and reagents

- Water-bath at 42 °C
- 37 °C incubator
- Plastic Coplin jars
- DAPI (4′,6-diamidino-2-phenylindole): 2 μg/ml (Vector)
- Coverslips: 24 × 40 mm and 22 × 22 mm
- Antifade (Vector)
- 2 × SSC
- Propidium iodide (PI) (Vector)

Avidin–biotin system

- Wash solution (ST): 4 × SSC, 0.1% Tween 20
- Blocking solution: 4 × SSC, 3% BSA, 0.1% Tween 20
- Antibody diluent buffer (SBT): 4 × SSC, 1% BSA, 0.1% Tween 20
- FITC–avidin: 5 μg/ml in SBT (Vector, Sigma)
- Biotinylated goat anti-avidin antibody: 5 μg/ml in SBT (Roche Diagnostics, Vector)

Digoxigenin (DIG) system

- Wash solution (TNT): 0.1 M Tris–HCl, 0.15 M NaCl, 0.05% Tween 20
- Blocking solution (TNB): 0.1 M Tris–HCl, 0.15 M NaCl, 0.5% Boehringer blocking reagent pH 7.5 (Roche Diagnostics)
- Mouse anti-DIG antibody: 0.5 μg/ml in TNB (Roche Diagnostics)
- DIG labelled anti-mouse antibody: 2 μg/ml in TNB (Roche Diagnostics)
- Anti-DIG–FITC or rhodamine: 2 μg/ml in TNB (Roche Diagnostics)

A. Avidin/biotin system

1 Drain excess fluid from the edge of slides following the last stringency wash.

2 Add 40 μl of the blocking solution to each slide, cover with a 24 × 40 mm coverslip, and incubate at 37 °C for 20 min.

3 Tap the coverslips off gently on the edge of a plastic beaker. Drain the excess blocking reagent. DO NOT RINSE.

4 Add 40 μl FITC–avidin to each slide,[a] cover with a coverslip, and incubate as in step 2.

5 Remove the coverslips as in step 3, or allow them to slide off in the first wash.

6 Wash the slides for 5 min in each of three Coplin jars containing pre-warmed ST (45 °C) with gentle shaking.

7 If amplification is required, continue; otherwise go to part C.

8 Add 40 μl of biotinylated goat anti-avidin antibody, cover with a coverslip, incubate for 20 min at 37 °C, and proceed as in steps 5 and 6.

9 Repeat steps 4–6 and counterstain.

B. DIG system

1 Follow part A (avidin/biotin detection) substituting as follows:

(a) TNB blocking solution for blocking solution in step 2.

(b) Mouse anti-DIG for FITC–avidin in step 4.

(c) TNT for ST wash in step 6.

(d) DIG–anti-mouse antibody for anti-avidin antibody in step 8.[b]

(e) FITC–anti-DIG antibody for the second addition of FITC–avidin in step 9.

C. Counterstain

1 Counterstain slides with DAPI (and/or PI[c]) in antifade. Add ~15 μl per slide and cover with a 22 × 22 mm coverslip. Store in a dry, light-protected slide container at 4 °C.

[a] As fluorescent compounds are sensitive to light, the slides should be kept covered following the first application of the fluorescent–antibody complex. If possible the remaining steps should be performed away from direct exposure to light or be done in a designated dark room area. Keep slides moist at all times.

[b] When detecting two or more probes simultaneously, the choice of antibody and/or the order of application is critical in order to avoid species cross-reaction, i.e. if detecting a biotinylated probe with a goat anti-avidin antibody, do not use a sheep anti-mouse antibody to detect a DIG labelled probe, due to the species similarity and cross-reaction between sheep and goat.

[c] PI counterstain does not produce a banding pattern on chromosomes, and therefore has limited use in mapping studies. It is used more frequently in clinical FISH (see Chapter 9).

8 Chromosome counterstaining and banding

To successfully map a DNA probe by FISH, both the specific chromosome and the banding localization must be identified. Although chromosomes can be G-banded prior to or following FISH probe localization, one cannot view the probe signal and banding at the same time. The two most common methodologies for simultaneous detection of hybridized probe and banding are R-banding (23) and DAPI-banding (24). R-banding is produced when synchronized cells incorporate BrdU during release from the cell cycle block (see *Protocol 3*). Ultraviolet irradiation of chromosomes counterstained with PI, in the presence of Hoechst 33258 will produce an R-banding pattern. The chromosome-specific banding patterns are formed as PI is taken up preferentially by the GC-rich regions. Chromomycin A3 and distamycin A have also been used successfully to produce R-banding suitable for FISH mapping (25). However, R-banding is not appropriate for mapping probes when one of the fluorochromes used is Texas Red or TRITC (rhodamine).

DAPI-banding, or reverse G-banding, is the more common method for mapping, especially for relational mapping of two or more probes. This counterstain is taken up preferentially by double-stranded AT-rich sequences, producing characteristic light and dark banding patterns specific for each chromosome. DAPI has the advantages that it can be used with both FITC and TRITC labelled probes and with target DNA from synchronized or non-synchronized cells. The disadvantage of DAPI is that it requires a UV light source for excitation (350 nm) and visualization. When counterstaining with DAPI or PI, care must be taken not to obscure the banding pattern or signal (PI and FITC) with too strong a counterstain.

9 Microscopy and image analysis

The basic requirement for visualizing fluorescently labelled probes is a high quality epifluorescence microscope (Olympus, Zeiss, Nikon, Leitz, etc.) equipped with a properly aligned high intensity mercury arc lamp light source (100 watt bulb), and objectives that contain lenses of high numerical aperture and low self-fluorescing UV transmitting glass. It is useful to have a range of objective magnifications—dry $\times 10$ and $\times 20$ for scanning metaphases and nuclei, and oil $\times 63$ and $\times 100$ for visualization and localization of the signal. The oil used for the high power objectives should be non-fluorescing. Mixing oils from different sources should be avoided as this can cause cloudiness and hinder signal visualization.

Filter sets (Chroma or Omega Optical) chosen for viewing fluorescence signals and counterstained chromosomes will depend on which fluorochromes are being used and the type of mapping being performed. Narrow bandpass filters have optimized peak excitation and emission wavelengths for a single fluoro-chrome such as DAPI or FITC to provide maximum signal intensity. Wide band-pass filters are used to view two (dual bandpass) or three (triple bandpass) colours simultaneously, and may result in some reduction of signal intensity due to the 'bleed-through' of emission signals (see Chapter 2). Since direct visualiza-tion of signals from small single copy or unique sequence probes is often diffi-cult, even at high magnification ($\times 100$), the use of a good quality digital camera such as a cooled charged-couple device (CCD) camera is recommended for fine mapping studies. Cooling the camera to $-38\,°C$ reduces 'dark current' noise and minimizes background signals caused by the camera itself. Most high resolution CCD cameras are cooled monochrome CCD cameras which capture greyscale images based on light intensity. Greyscale images are captured separately for each fluorochrome and transferred into an image analysis software program such as Adobe Photoshop for addition of colour and merging. The captured images are easily enhanced (cropped, labelled, background removed) and archived. However, registration is a critical issue when capturing images using two different filters. Great care must be taken when moving from one filter (DAPI) to another (FITC) as slight changes in registration can potentially result in an error in banding localization. Automatic computer controlled filter wheels and software packages that enable the capture of real-time images now available can minimize this problem.

10 FISH mapping points to consider

10.1 FISH mapping of single probes to metaphase chromosomes

10.1.1 Genomic probes

A key factor in FISH mapping is the identification of a true signal. The appearance of two pairs of doublet signals at the same chromosomal location in a metaphase spread is the hallmark of true gene localization (*Figure 2B*; see also *Plate 1B*). Hybridization efficiency (% metaphases showing two pairs of consistent doublet signals) determines the number of metaphases that need to be scored to establish chromosomal localization and can vary with the type and size of the probe. Larger genomic probes (PACs, BACs, YACs, and most cosmids), usually show a high (~90%) hybridization efficiency and require analysis of only 15–20 metaphases. Smaller genomic plasmid probes (<10 kb) usually demonstrate a lower hybridization efficiency (<20%) and require analysis of 25–50 metaphases. Occasionally, when mapping a genomic probe, a second signal (often weaker and lower frequency) may consistently appear at a different chromosomal location. This indicates the presence of another gene family member or a pseudogene and requires further molecular analysis to identify the nature of this second site (26, 27).

10.1.2 Unique sequence probes

Visualization of unique sequence probes (cDNAs) is dependent on the size of the target gene and its intron/exon structure. In general, cDNAs >1 kb can be visualized directly by routine FISH (*Figure 2A*; see also *Plate 1A*), but at low efficiency. Successful mapping of probes <1 kb is less frequent but has been reported (21) using a tyramide amplification system (19). An alternative approach that we have found useful is to screen a human genomic PAC or BAC library (see Section 2.1) with the cDNA in order to identify a corresponding genomic clone(s) that can then be easily mapped by FISH. If one genomic clone is identified, the mapping may result in:

(a) Localization to a single chromosome indicating the true mapping site.

(b) Localization to two different chromosomes, suggesting the presence of a double or chimeric clone and warrants further checking before continuing.

Often several PAC/BAC clones will be identified for one cDNA. These clones should be first checked by PCR to verify the correct insert sequence prior to proceeding with the mapping experiments. If two or more 'verified' clones are identified, the mapping may result in:

(a) Localization to the same chromosomal site, indicating true mapping site and redundancy within the library.

(b) Localization to different chromosome sites, indicating the presence of a gene family or a pseudogene. This approach can lead to the identification and localization of novel gene family members and pseudogenes (27, 28).

Since PAC/BAC libraries are publicly available (see Section 2.1), this approach has the advantage of providing both genomic clones and localization data for expressed sequences.

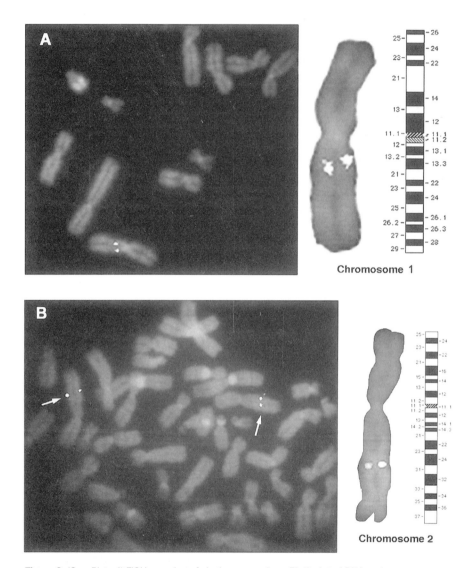

Figure 2 (See *Plate 1*) FISH mapping of single copy probes. Biotinylated DNA probes were hybridized to normal human metaphase chromosomes and detected with FITC. The chromosomes were counterstained with DAPI (see *Protocols 8–10*). Signals were visualized using a digital CCD camera (Photometrics). The band assignment was determined by analysing the banding pattern and measuring the fractional chromosome length. (A) Doublet signals of a 1.7 kb cDNA (unique sequence) probe are shown in a partial metaphase to localize to 1q13.3 (from ref. 9). (B) Doublet signals of a genomic PAC probe (~100 kb) are shown to map to 2q31 on both homologues of chromosome 2.

10.2 Relational mapping with multiple probes

10.2.1 Metaphase

Multicolour metaphase FISH enables the ordering and relative band positioning of two or more probes on a chromosome (29, 30). The number of different probes one can map simultaneously depends on the number of available fluorochromes, the microscope filter capabilities, the imaging software, and the physical distance between the probes. Due to the condensed nature of metaphase DNA, the level of resolution for two or more DNA probes on metaphase chromosomes is 1–2 Mb (31, 32). Probes can be ordered on metaphase chromo-

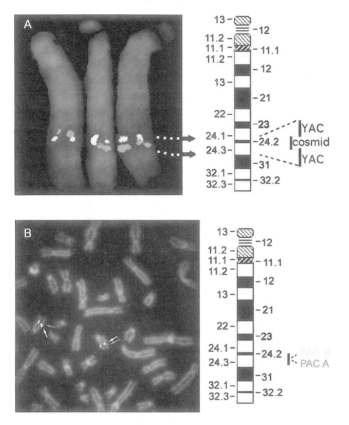

Figure 3 (See *Plate 2*) Relational mapping on metaphase chromosomes using two fluorochromes. (A) Three probes. Pairwise mapping was used to order two YACs and one cosmid probe within a YAC contig which mapped to 14q24.2-24.3. The two YACs were labelled with DIG or biotin and the cosmid with biotin. DIG was detected with rhodamine (red) and biotin with FITC (yellow). Left chromosome: DIG-YAC 1 plus cosmid; middle chromosome: DIG-YAC 2 plus cosmid; right chromosome: biotin-YAC 1 plus DIG-YAC 2. Thus the order shown is YAC 1–cosmid–YAC 2. Note: YAC 2 often demonstrated double hybridization signals in a lateral orientation suggesting two domains of strong hybridization or clustering of Alu sequences within the YAC (see right chromosome). (B) Two probes. Two PAC probes localized to 14q24-31 by FISH were labelled with different haptens (DIG-PAC A and biotin-PAC B) and hybridized simultaneously to normal human metaphase chromosomes counterstained with DAPI. PAC A (detected with rhodamine, red) localized telomeric to PAC B (detected with FITC, yellow).

somes relative to the telomere, centromere, a specific band, a breakpoint region, or a known genetic marker (*Figure 3*; see also *Plate 2*). Mapping probes relative to a specific chromosomal breakpoint can provide information regarding their biological and/or diagnostic significance (see Chapter 9). Identification of a gene that spans a breakpoint requires that the sequence length on either side of the

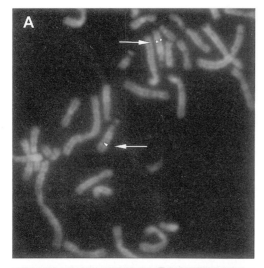

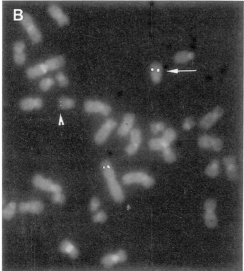

Figure 4 (See *Plate 3*) Mapping across a chromosome breakpoint. (A) A cosmid probe (14 kb) hybridized to normal human metaphase chromosomes showed doublet signals at 14q24 on both chromosome homologues. (B) Hybridization of the same cosmid probe to metaphase chromosomes from a patient with a 2:14 (p25;q24) translocation. The probe demonstrates doublet signals on both normal (arrow) and derivative (arrowhead) chromosome 14 as well a doublet signal on the short arm of derivative chromosome 2 (no arrow). A consistently stronger signal was noted on der2 indicating the breakpoint is located closer to the proximal (centromeric) end of the cosmid.

breakpoint to be of sufficient length to produce a visible signal (*Figure 4*; see also *Plate 3*). Mapping DNA probes relative to other chromosomal abnormalities, regions of amplification or deletion, or to tandem repeats can also be performed using metaphase chromosomes provided probes are available for the abnormal region. For applications of multicolour probe cocktails for mapping translocation breakpoints (SKY™) see Chapter 10.

10.2.2 Interphase

Higher resolution relational mapping can be obtained using interphase (G1–G0) nuclei where the DNA is not as contracted as that in metaphase chromosomes. In interphase nuclei, probes separated by 50–100 kb can be distinguished as separate signals (33). It is necessary to use probes with high hybridization efficiency, as interphase mapping is less efficient than metaphase mapping. The highest resolution relational mapping (1–40 kb) is achieved with free (34, 35) or stretched (36) chromatin (fibre FISH) and is described in Chapter 5. In interphase nuclei, a linear relationship between the genomic and the physical distance has been demonstrated over distances of about 50 kb–1 Mb (31, 37, 38) in certain regions of the genome. Since chromosome packing varies within the nucleus, any distance measurements should be verified with known markers before attempting to map unknown markers in a specific region. Although interphase FISH mapping is useful for probe ordering in the 50 kb–2 Mb range, actual distance measurements may be more readily obtained from genome database contig information (see also ref. 16).

In summary, the identification of clones and markers now available through various WWW sites together with FISH mapping can provide a powerful strategy for ordering and mapping novel genes and ESTs.

References

1. Beaulieu, M., Levesque, E., Tchernof, A., Beatty, B. G., Belanger, A., and Hum, D. W. (1997). *DNA Cell Biol.*, **16**, 1143.
2. Malik, N., Canfield, V., Sanchez-Watts, G., Watts, A. G., Scherer, S., Beatty, B. G., *et al.* (1998). *Mamm. Genome*, **9**, 136.
3. Beatty, B. G., Qi, S., Pienkowska, M., Herbrick, J. A., Scheidl, T., Zhang, Z. M., *et al.* (1999). *Genomics*, **62**, 529.
4. Burger, A. M., Zhang, X., Li, H., Ostrowski, J. L., Beatty, B., Venanzoni, M., *et al.* (1998). *Oncogene*, **16**, 2459.
5. Hannigan, G. E., Bayani, J., Weksberg, R., Beatty, B., Pandita, A., Dedhar, S., *et al.* (1997). *Genomics*, **42**, 177.
6. Crawford, M. J., Lanctot, C., Tremblay, J. J., Jenkins, N., Gilbert, D., Copeland, N., *et al.* (1997). *Mamm. Genome*, **8**, 841.
7. Pennica, D., Swanson, T. A., Shaw, K. J., Kuang, W. J., Gray, C. L., Beatty, B. G., *et al.* (1996). *Cytokine*, **8**, 183.
8. Baker, M. D., Read, L. R., Beatty, B. G., and Ng, P. (1996). *Mol. Cell. Biol.*, **16**, 7122.
9. Janicic, N., Soliman, E., Pausova, Z., Seldin, M. F., Riviere, M., Szpirer, J., *et al.* (1995). *Mamm. Genome*, **6**, 798.
10. Lichter, P., Cremer, T., Borden, J., Manuelidis, L., and Ward, D. C. (1988). *Hum. Genet.*, **80**, 224.

11. Brown, M. J. and Lawce, H. J. (1997). In *The AGT cytogenetics laboratory manual* (ed. M. J. Barch, T. Knudsen, and J. L. Spurbeck), p. 77. Lippincott–Raven, Philadelphia.

12. Takahashi, E., Hori, T., O'Connell, P., Leppert, M., and White, R. (1990). *Hum. Genet.*, **86**, 14.

13. Cherif, D., Julier, C., Delattre, O., Derre, J., Lathrop, G. M., and Berger, R. (1990). *Proc. Natl. Acad. Sci. USA*, **87**, 6639.

14. Gustashaw, K. M. (1997). In *The AGT cytogenetics laboratory manual* (ed. M. J. Barch, T. Knutsen, and J. L. Spurbeck), p. 259. Lippincott–Raven, Philadelphia.

15. Priest, J. H. (1997). In *The AGT cytogenetics laboratory manual* (ed. M. J. Barch, T. Knutsen, and J. L. Spurbeck), p. 173. Lippincott–Raven, Philadelphia.

16. Leversha, M. A. (1997). In *Genome mapping: a practical approach* (ed. P. H. Dear), p. 199. IRL Press, Oxford.

17. Wiegant, J., Wiesmeijer, C. C., Hoovers, J. M., Schuuring, E., d'Azzo, A., Vrolijk, J., *et al.* (1993). *Cytogenet. Cell Genet.*, **63**, 73.

18. Yu, H., Chao, J., Patek, D., Mujumdar, R., Mujumdar, S., and Waggoner, A. S. (1994). *Nucleic Acids Res.*, **22**, 3226.

19. Raap, A. K., van de Corput, M. P., Vervenne, R. A., van Gijlswijk, R. P., Tanke, H. J., and Wiegant, J. (1995). *Hum. Mol. Genet.*, **4**, 529.

20. Kerstens, H. M., Poddighe, P. J., and Hanselaar, A. G. (1995). *J. Histochem. Cytochem.*, **43**, 347.

21. Schriml, L. M., Padilla-Nash, H. M., Coleman, A., Moen, P., Nash, W. G., Menninger, J., *et al.* (1999). *Biotechniques*, **27**, 608.

22. Speel, E. J., Hopman, A. H., and Komminoth, P. (1999). *J. Histochem. Cytochem.*, **47**, 281.

23. Takahashi, E., Koyama, K., Hirai, M., Itoh, H., and Nakamura, Y. (1995). *Cytogenet. Cell Genet.*, **68**, 112.

24. Heng, H. H. and Tsui, L. C. (1993). *Chromosoma*, **102**, 325.

25. Korenberg, J. R. and Chen, X. N. (1995). *Cytogenet. Cell Genet.*, **69**, 196.

26. Malik, N., Canfield, V., Sanchez-Watts, G., Watts, A. G., Scherer, S., Beatty, B. G., *et al.* (1998). *Mamm. Genome*, **9**, 136.

27. Turgeon, D., Carrier, J. S., Levesque, E., Beatty, B. G., Belanger, A., and Hum, D. W. (2000). *J. Mol. Biol.*, **295**, 489.

28. Meloche, S., Gopalbhai, K., Beatty, B. G., Scherer, S. W., and Pellerin, J. (2000). *Cytogenet. Cell Genet.*, **88**, 249.

29. Ried, T., Baldini, A., Rand, T. C., and Ward, D. C. (1992). *Proc. Natl. Acad. Sci. USA*, **89**, 1388.

30. Heppell-Parton, A. C., Albertson, D. G., Fishpool, R., and Rabbitts, P. H. (1994). *Cytogenet. Cell Genet.*, **66**, 42.

31. Lawrence, J. B., Singer, R. H., and McNeil, J. A. (1990). *Science*, **249**, 928.

32. Trask, B., Fertitta, A., Christensen, M., Youngblom, J., Bergmann, A., Copeland, A., *et al.* (1993). *Genomics*, **15**, 133.

33. Lawrence, J. B. (1990). In *Genome analysis Volume 1: Genetic and physical mapping* (ed. K. E. Davies and S. M. Tilghman), p. 1. Cold Spring Harbor Laboratory Press, Cold Spring Harbor.

34. Heng, H. H., Squire, J., and Tsui, L. C. (1992). *Proc. Natl. Acad. Sci. USA*, **89**, 9509.

35. Florijn, R. J., Bonden, L. A., Vrolijk, H., Wiegant, J., Vaandrager, J. W., Baas, F., *et al.* (1995). *Hum. Mol. Genet.*, **4**, 831.

36. Parra, I. and Windle, B. (1993). *Nature Genet.*, **5**, 17.

37. Trask, B. J., Massa, H., Kenwrick, S., and Gitschier, J. (1991). *Am. J. Hum. Genet.*, **48**, 1.

38. Senger, G., Ragoussis, J., Trowsdale, J., and Sheer, D. (1993). *Cytogenet. Cell Genet.*, **64**, 49.

Chapter 4
Murine chromosome preparation

Sabine Mai
Manitoba Institute of Cell Biology, 675 McDermot Avenue, Room 6046, Winnipeg R3E 0V9, Canada.

Francis Wiener
Microbiology and Tumor Biology Center, Karolinska Institute, PO Box 280, S-17177 Stockholm, Sweden.

1 Murine chromosome preparation for banding and *in situ* hybridization procedures

1.1 Introduction

Cytogenetic analysis of mouse chromosomes demands high quality metaphase plates suitable for various banding procedures and *in situ* hybridization protocols. The methods of choice must yield chromosome preparations that meet the criteria of high quality plates, namely:

(a) Uniformly spread chromosomes within the limits of the high power visual field of the microscope.

(b) Well preserved chromosome morphology.

(c) Absence of overlapping chromosomes (1, 2).

This requirement is underscored by the fact that, unlike the human karyotype where the identification of the 23 chromosome homologues is facilitated by their sizes and arm ratio differences, the analysis of mouse chromosomes is more cumbersome due to the uniformity of the mouse karyotype which is composed of 20 homologue pairs of an exclusively acrocentric type (3).

This chapter describes methods of mouse chromosome preparation used in our laboratory which were found to be reliable and yielding consistent results irrespective of the tissue source. The most commonly utilized tissues for chromosome preparation are mouse bone marrow, spleen, lymph nodes, and thymus because each provides a sufficient number of dividing cells. The same methods can be applied for the analysis of mouse tumours such as murine thymic and/or splenic lymphoma and mouse plasmacytoma. We have preferentially used the 'direct' preparation method; i.e. chromosome preparations in which the cells have not been propagated *in vitro* (1, 4, 5).

The common steps in chromosome preparation of both normal and tumour tissue are:

(a) Collection of the mitotic cells.

(b) Swelling of the cells by hypotonic solution.

(c) Fixation of the hypotonized cells.

(d) Spreading the chromosomes on a slide by appropriate methods (6, 7).

While variation in steps (a), (b), and (d) do not conclusively influence the quality of the chromosome preparation, the fixation procedure (step c) has a critical impact on whether the chromosome preparation will match the criteria of high quality metaphase plates.

The fixation has two aims:

(a) A gradual dehydration of the hypotonized cells.

(b) Preservation of the chromosome morphology.

This process must be conducted in such a way that both the shrinkage of the hypotonized cell and the disruption of chromosome morphology can be avoided. Inappropriate fixation will contract the cells, and the chromosomes will crowd within the cell membrane minimizing the likelihood of an adequate spreading. Figuratively, a perfectly performed fixation transforms the hypotonized, swollen mitotic cell into a fragile crystal ball which will easily release the chromosomes when its membrane disintegrates during the spreading procedure.

There are three additional parameters which may influence the quality of chromosome preparation.

(a) Standard chromosome preparation protocols frequently recommend the use of different concentrations of colcemid both for accumulation of a higher number of mitotic cells and for the disruption of the mitotic spindle. Though the use of colcemid does have the outcome mentioned above, its application has predominately an unfavourable effect on the chromosome morphology. Contraction of mitotic chromosomes and the consequent reduction in their size due to prolonged colcemid treatment of the cells make them less suitable for any banding procedures, and may result in a low quality of G-banding.

(b) In the case of pellet fixation (*Protocol 2*), the number of cells per centrifuge tube may also influence the quality of the metaphase plates. The rapid penetration of the fixative depends on the thickness of the cell pellet. An incomplete penetration due to a thick pellet (usually more than 1 mm) hinders the gradual and uniform dehydration of the hypotonized cells with negative consequences, such as cell shrinkage and impaired spreading of chromosomes, resulting in a high number of overlapping chromosomes.

(c) Both normal tissues such as bone marrow, spleen, and tumour tissues contain red blood cells (RBCs) which interfere with the fixation process of the chromosome preparation. It is advisable to remove the RBCs from the initial cell suspension. There are a number of methods for lysing the RBCs; we recommend treating the cell suspension with ACK solution (see *Protocol 1*).

Protocol 1

Chromosome preparations from mouse bone marrow cells

Equipment and reagents

- Centrifuge tubes 15 ml (Fisher, 05-539-1)
- 3 ml and 5 ml syringe (Becton Dickinson, 309585, 309603)
- 25 and 20 gauge needles (Becton Dickinson, 305122, 305175)
- Pasteur pipettes (Fisher, 13-678-20D)
- Scissors and forceps (VWR, 25874-105, 25601-146)
- 40 μm nylon mesh
- Petri dishes 5 ml and 10 ml (Invitrogen, Nunc, 150288, 150380)
- ACK solution: 8.28 g NH_4Cl, 1 g $KHCO_3$, 0.04 g EDTA in 1000 ml distilled water pH 7.4 (Sigma, A4514, P4913, E9884)
- RPMI 1640 medium (Canadian Life Technologies, 31800-089) with and without 10% FCS supplementation (Invitrogen, 10437-028)

Method

1 Flush out the bone marrow plug from both femurs into a 15 ml centrifuge tube with 3 ml cold RPMI 1640 using a 3 ml syringe fitted with a 25 gauge needle.

2 Disrupt the bone marrow clumps first by pipetting followed by repeated syringing (three to four times) through a 20 gauge needle.

3 Centrifuge at 200 g for 5 min.

4 To lyse the red blood cells (RBCs), resuspend the cell pellet by repeated pipetting for 1–2 min in 1–2 ml cold ACK solution.

5 Centrifuge at 200 g for 5 min.

6 Decant the supernatant and resuspend the cell pellet in 2–3 ml cold RMPI 1640.

7 Centrifuge at 200 g for 5 min.

8 Decant the supernatant and transfer the cell suspension into a Petri dish in 10 ml RPMI 1640 medium supplemented with 10% FCS.

9 Incubate the cell suspension at 37 °C for 30 min.

10 Aliquot the cell suspension into 15 ml centrifuge tubes (approx. 5–8 × 10^6 cells/ tube).

11 Centrifuge at 200 g for 5 min, discard the supernatant, and follow one of the fixation procedures described in *Protocols 2, 3a,* or *3b*.

Protocol 2

Pellet fixation[a]

Equipment and reagents

- See *Protocol 1*
- 0.075 M KCl hypotonic solution (KCl: Sigma, P3911)
- Fresh fixative solution: 3 vol. methanol (Fluka):1 vol. acetic acid (Fisher, A38S-225)

Method

1 Add to the cell pellet five to ten drops of 0.075 M KCl hypotonic solution. Dislodge the pellet by gently flicking the centrifuge tube with a finger. Slowly add more hypotonic solution up to 5 ml. Mix the cells with a Pasteur pipette.

2 Keep the cells at room temperature for 10–15 min.[b]

3 Centrifuge at 200 g for 5 min.

4 Discard the supernatant without disturbing the cell pellet.

5 Layer two drops of fixative on to the surface of the cell pellet, rolling it along the side of the tube, for 1 min. Continue by carefully adding three, four, five drops of fixative at 1 min intervals, followed by 10, 15, 20, and 30 drops at 2 min intervals without disturbing the cell pellet.

6 Discard the supernatant and resuspend the cell pellet in 2 ml fresh fixative by repeated pipetting until a cell suspension free of cell clumps is formed.

7 Gradually add more fixative to a final volume of 5 ml. Shake the tube while adding the fixative. If necessary, remove the cell clumps by filtrating it through 40 μm nylon mesh.

8 Centrifuge at 200 g for 5 min.

9 Discard the supernatant and slowly add 5 ml fresh fixative. Allow to stand at room temperature for 10 min. Centrifuge as in step 8.

10 Discard the supernatant and resuspend the cells in 1 ml of fresh fixative.

11 Repeat steps 8–10 twice but let the re-fixed cell pellet stand between each step for 20 and 30 min, respectively.

12 Resuspend the cells in 1 ml of fixative and drop two or three drops on a wet slide chilled in ice-cold distilled water. Air dry the slides at room temperature.

[a] Pellet fixation results in associated chromatids suitable for different (Q-, G-, and R-) banding procedures.

[b] At this step it is advisable to ascertain whether the mitotic index is >1% (*Protocol 8*).

Protocol 3a

Suspension fixation[a]

Equipment and reagents

- See *Protocol 2*

Method

1 Proceed according to *Protocol 1*, steps 1–7.

2 Layer 10–15 drops of 0.075 M KCl hypotonic solution on the cell pellet and dislodge the pellet by gently flicking the centrifuge tube with a finger. Slowly add more hypotonic solution to a final volume of 1 ml. Mix the cells with a Pasteur pipette until a uniform cell suspension is formed.

3 Add two drops of fixative and gently shake the cell suspension for 30–60 sec.

4 Add in continuation 4, 6, 8, 10, 15, and 20 drops of fixative at 1 min intervals while shaking the cell suspension.

5 Slowly add additional fixative up to a total volume of 5 ml. Agitate with a Pasteur pipette until a uniform cell suspension is formed.

6 Centrifuge at 200 g for 5 min.

7 Discard the supernatant and slowly add new fixative up to a total volume of 5 ml. Allow to stand for 15 min. Centrifuge as in step 6.

8 Discard the supernatant and resuspend the cell pellet in 1 ml fixative.

9 Repeat steps 6–8 twice.

10 Resuspend the cells in 1 ml fixative and drop two or three drops on a wet slide chilled in ice-cold distilled water. Air dry the slides at room temperature.

[a] Suspension fixation results in detached chromatids that are more appropriate for different *in situ* hybridization approaches.

Protocol 3b

Alternative suspension fixation[a]

Equipment and reagents

- See *Protocol 2*

Method

1 Add 5 ml of 0.075 M KCl hypotonic solution to the cell pellet and mix the cells until a uniform cell suspension is formed. Incubate at room temperature for 30 min.

2 Slowly layer 1 ml of fresh fixative on top of the 5 ml hypotonic solution. Invert tube three or four times.

Protocol 3b continued

3   Centrifuge at 200 g for 5 min.

4   Discard supernatant and resuspend the cell pellet in 5 ml fixative.

5   Repeat steps 3 and 4 twice.

6   Resuspend the cell pellet in 5 ml fixative and drop cells on slides as described in *Protocol 2*.

[a] Cells fixed according to the above protocol can be kept at 4 °C and can be reused up to five years. The fixative must be replaced once a week.

Protocol 4

Chromosome preparation from mouse splenic cells

Equipment and reagents

• See *Protocol 1*

Method

1   Remove the whole spleen and place into a Petri dish containing 10–15 ml cold RMPI 1640.

2   Grasp the spleen firmly with a forceps and perforate the spleen capsule in different locations with an injection needle (27 gauge).

3   Flush out the cells with a 5 ml syringe, fitted with a 20 gauge needle, containing 5 ml cold RMPI 1640, which is inserted into the upper end of the spleen.

4   Repeat flushing until the desired number of cells is obtained (approx. 6–7 $\times$ 10^7).

5   Follow *Protocol 1*, steps 3–11.

Protocol 5

Chromosome preparation from mouse thymic (and lymph node) cells[a]

Equipment and reagents

• See *Protocol 1*

• Colcemid (Invitrogen, 15210-040)

Method

1   Remove the whole thymus (or axillar and inguinal lymph nodes) and place into a Petri dish containing 10–15 ml cold RMPI 1640 on ice.

Protocol 5 continued

2 Insert a 3 ml syringe with a 20 gauge needle into the thymic lobes (or lymph nodes) and flush out the cells with cold RMPI 1640.

3 Repeat flushing until the desired number of cells is obtained (approx. 3–5 $\times$ 10^7 cells).

4 Follow *Protocol 1*, steps 3–11, and perform one of the pellet or suspension fixation procedures outlined in *Protocols 2* and *3*.

[a] If the mitotic index is low (<1%) add colcemid to the single cell suspension at a final concentration of 0.03 µg/ml at *Protocol 1*, step 9.

Protocol 6

Chromosome preparation of mouse plasmacytoma[a]

Equipment and reagents

- See *Protocol 1*

Method

1 Centrifuge approx. 10–15 ml of aspirated ascitic fluid at 200 g for 5 min.

2 Discard the supernatant and resuspend the pellet by repeated pipetting for 1–2 min in 5 ml cold ACK solution.

3 Centrifuge at 200 g for 5 min.

4 Decant the supernatant and thoroughly resuspend the cell pellet in 2–3 ml cold RPMI 1640.

5 Centrifuge at 200 g for 5 min.

6 Decant the supernatant, resuspend the cells, and transfer the cell suspension to a Petri dish containing 10 ml RPMI 1640 medium and 10% FCS.

7 Incubate the cell suspension at 37 °C for 30 min.

8 Collect the supernatant and centrifuge at 200 g for 5 min.

9 Discard the supernatant without disturbing the cell pellet.

10 Follow either the pellet or suspension fixation procedures outlined in *Protocols 2* and *3*.

[a] Mouse plasmacytoma develops in the peritoneal cavity of the mouse as an ascitic type tumour (8). In addition to the plasmacytoma cells the ascitic fluid contains a low number of lymphocytes as well as a considerable number of red blood cells (RBCs) and macrophages. We recommend the lysis of RBCs as described in *Protocol 1*. A practical method for removal of the macrophages is to exploit their property of adherence to plastic surfaces (this protocol, step 7). By using these approaches, the cell suspension will contain primarily plasmacytoma cells and only a low number of lymphocytes.

Protocol 7

Preparation of mouse chromosome spreads for molecular cytogenetics

Equipment and reagents

- See *Protocol 1*
- Microscope slides (O. Kindler GmbH, K102)
- Coverslips (O. Kindler GmbH)
- Slide warmer

Method

1   Place slide on dry ice until signs of freezing are visible (10–20 sec).

2   Place one to three drops of the fixed cell suspension[a] on the slide and place it immediately on to a 37 °C slide warmer until fixative begins to evaporate.[b]

3   Dip slide for 5 sec into 50% acetic acid solution.[c]

4   Place slide back onto the 37 °C slide warmer and wait until dry.

5   The slide is now ready for fluorescent *in situ* hybridization (FISH), chromosome painting, spectral karyotyping (SKY), M-FISH (multicolour FISH), or comparative genomic hybridization (CGH).[d,e]

[a] If fixed cell suspension has been prepared some time ago (>seven days), pellet cells and resuspend in fresh fixative.

[b] The optimal time of evaporation for the best chromosome spreading has to be established empirically.

[c] The solution can be kept for one week.

[d] Hybridization results will be best with slides that have been prepared one to three days prior to analysis.

[e] Evaluate the metaphase spreads by Giemsa staining (as described in *Protocol 9*) and mark with a diamond pen the best area containing high quality metaphases to be used for hybridization.

Protocol 8

Determination of the mitotic index of cell populations used for chromosome preparations

Equipment and reagents

- Microscope slides (O. Kindler GmbH, K102)
- Coverslips (O. Kindler GmbH)
- Stock staining solution: 1 g of crystal violet (Sigma, C-3886) dissolved in a mixture of 90 ml distilled water and 10 ml of fixative (3 vol. methanol: 1 vol. acetic acid)
- Pasteur pipettes
- Microcentrifuge tubes (Sarstedt, 72.690)
- Working solution: mix 1 vol. of the stock staining solution with 5 vol. distilled water; filter if necessary

Method

1   Place two drops of the hypotonized cell suspension and two drops of the working solution into a microcentrifuge tube.

2   Mix by shaking the tube.

3   Place two drops of the mixture onto a microscope slide and cover with a coverslip.

4   Determine the number of mitotic figures by counting 500 cells under the microscope (×20 objective).

2 Giemsa–trypsin banding of mouse chromosomes

Chromosomes start to de-condense during replication at the end of the S phase and become visible in the early prophase after the breakdown of the nuclear membrane (9–11). By the end of the prophase and early metaphase the chromosomes are de-spiralized (12). They are elongated and display a large number of bands and sub-bands (13). These chromosomes are the subject of high resolution banding analyses. A discussion of this latter approach is beyond the scope of this chapter.

During the ensuing stage in mid-metaphase the many sub-bands fuse together to form one band. The resulting banding pattern, for which the location, size, and staining intensity of each band is characteristic for each individual chromosome (14), is maintained during the whole mid-metaphase stage. Hence, the chromosomes of the mid-metaphase plates represent the genuine target in which the trypsin treatment of the late replicating chromatin will generate the specific G-band pattern (15, 16) required for their identification, according to the criteria of the International Committee on Standardized Genetic Nomenclature for Mice (2) see *Figure 1*.

The parameters with a decisive impact on the quality of the G-banding are as follows:

(a) The mitotic index of the cell population.

(b) The ageing of the slides.

(c) The length of the trypsin treatment.

1   A non-synchronized cell population with a mitotic index between 1–3% contains only a low percentage of metaphase plates suitable for G-banding. In our experience, the number of mid-metaphases on a slide containing spreads from non-synchronized cell populations may vary between three to six plates per slide. Hence, successful G-banding requires not only high quality chromosome preparations but also a cell suspension containing a relatively high number of mid-metaphase plates.

2   The critical impact of slide ageing that will result in a flawless G-banding is underscored by the well known fact that, unlike the human chromosomes, the G-band quality of the mouse chromosomes of a fresh preparation is so fuzzy that it makes their identification difficult if not impossible. Nevertheless, the quality of the banding improves continuously with ageing. A prolonged ageing however, makes the chromosomes resistant to trypsin digestion, with the result that the chromosomes will be uniformly stained without displaying any banding pattern.

The ageing (ripening) process is time- and temperature-dependent. In our experience, the optimum time period for ageing at room temperature is between two to four weeks. However, at 37 °C, the ripening process can be reduced to 10–14 days. The optimum ripening time is determined empirically and will vary with the source of the cells and type of preparation (*in vitro* culture or direct preparation).

The ageing process can be overcome by pre-treating the slides with a solution of H_2O_2 (*Protocol 10*) (17). By applying this approach, the chromosomes can be G-banded immediately after spreading.

3 The concentration of the trypsin solution and the length of the trypsin treatment are crucial in order to obtain high quality banding. We recommend using a low concentration of trypsin solution (*Protocol 9*) and always at a constant temperature. Note that the consistency of the banding is already adversely affected by minor changes in the concentration of the trypsin solution and its temperature. Since there are slight differences between each batch of trypsin samples provided by the same manufacturer, the appropriate concentration of the trypsin solution and the length of the treatment must be evaluated separately for each batch of trypsin.

Protocol 9

Giemsa–trypsin banding of mouse chromosomes (see *Figure 1*)

Reagents

- Aliquots of 1000 µl (1 mg/ml) trypsin solution (Invitrogen, 17068-024) kept at –20 °C
- Giemsa stock solution (Sigma, G-3032)
- Trypsin buffer: 0.15 M phosphate buffer pH 8; $K_2(Na)PO_4$ and $K(Na)H_2PO_4$

- Trypsin working solution: 10 µl stock solution/ml trypsin buffer (can be kept in Coplin jar at 4 °C for seven days)
- Freshly prepared Giemsa working solution: 250 µl Giemsa/ml phosphate (trypsin) buffer

Method

1 Immerse a slide into the Coplin jar with trypsin working solution for 10–60 sec.[a]

2 Rinse with tap-water.

3 Stain with Giemsa solution for 1–2 min.[b]

4 Rinse by a strong jet of tap-water to remove the stain precipitation that eventually sticks on to the slide.

5 Air dry.

[a] The appropriate duration of trypsin treatment is established empirically by checking the banding quality of the chromosome spreads obtained at different time intervals of trypsin digestion.

[b] Since the G-band staining efficiency of the Giemsa solution decreases rapidly it has to be freshly prepared.

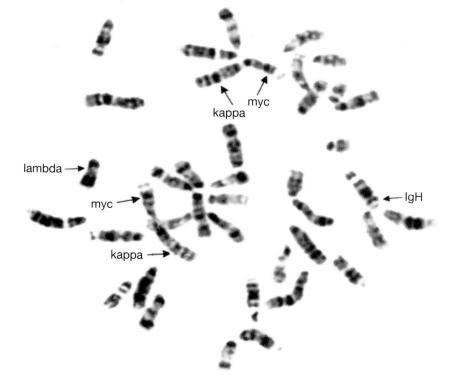

Figure 1 G-banded metaphase plate of translocation negative ABPC IL-10 plasmacytoma, generation 0. Arrows point to immunoglobulin loci (*IgH*, *lambda*, *kappa*) and to the *c-myc* locus. G-banding was carried out as described in *Protocol 9*.

Protocol 10

Ageing the slides by H₂O₂ pre-treatment

Reagents
- 3% H_2O_2 (Sigma, H-1009) working solution (can be stored in a Coplin jar at 4 °C for seven days)

Method

1 Immerse the freshly prepared slides into the H_2O_2 working solution for 30 min at room temperature.

2 Rinse with tap-water.

3 Air dry.

4 Proceed according to *Protocol 9*.

3 Molecular cytogenetic approaches for murine chromosomes

Classical cytogenetics of murine chromosomes has allowed cytogeneticists to define and study numerical and structural chromosomal aberrations in as much as they could be resolved in individual chromosome bands. Molecular cytogenetics has added additional information and refinement to classical cytogenetics. The new molecular cytogenetic approaches allow one to detect 'crytpic' chromosomal aberrations, to identify specific genes, and to examine marker chromosomes for their genomic content.

This section focuses on methods for FISH (see *Figure 2*), chromosome painting (see *Figure 3*), and SKY applied to mouse chromosomes (see *Figure 4*) as described in the protocols outlined below. Successful experimental results rely on the following individual parameters:

(a) Quality of the sample that is chosen for hybridization.

(b) Age of the slides.

(c) Purification of the hybridization probe, labelling of the probe, assessment of the labelling quality of the probe in comparison to a standard, and titration of each new probe batch in comparison with the old batch.

(d) The appropriate concentration of pepsin.

(e) The correct denaturation protocol.

1 Successful FISH, chromosome painting, and SKY experiments begin at the stage of sample preparation. As previously mentioned, well-spread chromosome plates lacking chromosome overlaps are an essential requirement for optimal and unequivocal results. The morphology of the mouse chromosomes as well as their sizes are critical factors if one wishes to perform chromosome banding after FISH, chromosome painting, or SKY analyses. *Protocol 17* outlines the recommended banding procedure after denaturation, hybridization, and washes.

2 The second important feature is the age of the slides. In our experience, in contrast to banding, the fresher the slides, the better the hybridization results. Fresh slides will yield higher hybridization efficiencies and brighter signals. In general, we perform FISH and chromosome painting the day of the slide preparation or the following day, and SKY within the first three days.

3 The purification of the probe for FISH is crucial. In our experience, gel- and membrane-purified probes work best. Each probe is re-checked after purification to confirm its size and check for the absence of DNA degradation.

Our laboratory generally prefers the indirect labelling of probes, since the amplification of the hybridization signals by specific antibodies (*Protocol 14*) allows for improved hybridization sensitivity. Direct labelling is, of course, feasible, but we recommend it only for probes exceeding 10 kb in size (i.e. cosmids, BACs, YACs).

The labelling of the probe and the assessment of the degree of labelling will determine the successful outcome of all subsequent FISH experiments. It is strongly recommended that each newly labelled probe be compared to the standard probe supplied by the manufacturer. Additionally, each new probe should be titrated by FISH against the same probe used in previous experiments. Such an experiment will tell whether the amount of probe used from the newly prepared batch is sufficient or should be increased or decreased. Once titrated, a new probe can be used routinely for many hybridization reactions. For example, 3 μg of DIG labelled *DHFR* probe (18) will last for about 100 hybridization reactions on metaphase chromosomes (19, 20).

It is important to note though, that the change of probe (or the new preparation of a probe) is not the sole factor determining the quality of the FISH hybridization. The FISH chromosome target preparation (interphase cells, tissue section, imprint, etc.) is of equally critical importance. Each batch of newly prepared chromosomes (interphase cells, tissue section, imprint, etc.) may require some adjustment of the amount of probe used in the hybridization reaction. Thus, successful FISH relies on the perfect match of the target material (chromosomes, interphase nuclei, sections, imprints) and the hybridization probe.

Two further steps have an important impact on successful FISH hybridizations; the treatment with pepsin, and the denaturation of the slides.

4 The concentration of pepsin given in *Protocol 13* is the average value around which one has to assess the optimal concentration for the individual chromosome preparations. Too much pepsin will adversely affect the morphology of the chromosomes. They will lose their structure and become blown up, fuzzy, and inappropriate for banding. However, chromosome painting and FISH are still possible, though the quality will be poor. If the pepsin concentration is too low, the chromosome morphology will be maintained, but the hybridization efficiency will be low. Therefore, one has to empirically determine the optimum pepsin concentration for each sample. As a general rule, we find that primary cells and immortalized cells require concentrations close to the given mean value of 50 μg/ml (*Protocol 13*), while highly transformed cells usually require much less. The pepsin concentration can be as low as 5 μg/ml. In some cases, the pepsin step (*Protocol 13*) can even be omitted.

5 The denaturation of the slides should be performed at 70 °C in 70% deionized formamide (Fluka)/2 × SSC pH 7.0, for 2 min. Over-denaturation of the chromosomes will have the same adverse effects on chromosome banding as over digestion with pepsin.

Protocol 11

Labelling of probes for FISH, assessment of labelling efficiency, and titration of probes[a]

Reagents

- 10 × hexanucleotide mix (Roche, 1277 081 or from DIG-DNA labelling Kit Roche, 1175 033)

- 10 × dNTP-mix (DIG): DIG-DNA labelling mix (Roche, 1277 065 or from DIG-DNA labelling kit) or

- Klenow enzyme (Roche, 1008 404 or from DIG-DNA labelling kit)

- Hybridization solution: 10% dextran sulfate in FSP (stock: 20% dextran sulfate in FSP)

- FSP: 50% formamide (deionized) (Fluka), 2 × SSC, 50 mM phosphate pH 7

- Buffer 1: 0.1 M maleic acid, 0.15 M NaCl pH 7.5 (1 litre: 100 ml of 1 M maleic acid, 30 ml of 5 M NaCl, approx. 50 ml of 5 M NaOH)

- Buffer 2: 1% blocking reagent (Nucleic Acid Detection Kit) in buffer 1 (10 × stock solution: 10% blocking reagent in buffer 1, dissolve in microwave, autoclave, store at 4 °C)

- dNTP-mix (biotin):

dATP	1 mM	1 μl 100 mM	(Roche, 1277 049)
dCTP	1 mM	1 μl 100 mM	(Roche, 1277 049)
dGTP	1 mM	1 μl 100 mM	(Roche, 1277 049)
dTTP	0.65 mM	0.65 μl 100 mM	(Roche, 1277 049)
Biotin-dUTP	0.35 mM	35 μl 1 mM	(Roche, 1277 049)
H_2O	61.35 μl; store at –20 °C		

- DNA dilution buffer: SS-DNA 50 μg/ml TE (10 mM Tris, 1 mM EDTA pH 8) (Roche, Nucleic Acid Detection Kit, 175 041)

- Anti-DIG–AP antibody: polyclonal sheep anti-DIG–Fab fragment-AP (Nucleic Acid Detection Kit) or

- Streptavidin–AP antibody (Roche, 1089 161)

- Buffer 3: 0.1 M Tris, 0.1 M NaCl, 0.05 M $MgCl_2$ pH 9.5 (1 litre: 12.1 g Tris base, 5.84 g NaCl, 10.17 g $MgCl_2x6H_2O$)

- Substrate solution (fresh): 5 ml buffer 3 + 22.5 μl NBT (nitroblue tetrazolium salt) solution + 17.5 μl X-phosphate solution (Nucleic Acid Detection Kit)

- TE: 10 mM Tris, 1 mM EDTA pH 8

- DEAE NA45 paper (Schleicher & Schuell, 417069)

- 4 M LiCl

- Absolute and 70% EtOH

Method

1 Gel purify the genomic fragments or cDNAs of interest using DEAE NA45 paper and elute DNA overnight in 1 M NaCl at 68 °C.

2 Precipitate the DNA following the addition of 2.5 vol. of cold 90% EtOH and resuspend it in a small volume (20–50 μl) of TE. An aliquot of each DNA fragment should be re-examined on a 2% agarose gel. If the size of the fragment is correct and the DNA is intact, proceed to label the probe.

3 Prior to the labelling reaction, denature 3 μg of probe in 15 μl H_2O by boiling for 10 min and then put on ice for 1–2 min.

Protocol 11 continued

4 Label the probe overnight at 37 °C by adding 2 ml of hexanucleotide mixture, 2 μl dNTP-mixture, and 1 μl Klenow.[a,b]

5 Stop the reaction with 2 μl of 0.2 M EDTA and precipitate the probe using 2.5 μl of 4 M LiCl and 75 μl EtOH (–20 °C) for 30 min at –80 °C or 2 h –20 °C.

6 Centrifuge the probe at 3750 g for 15 min at 4 °C.

7 Wash the pellet with 70% EtOH (–20 °C) and centrifuge at 3750 g for 15 min at 4 °C.

8 Dry the pellet for 5 min in a Speed Vac and then dissolve the pellet in 50 μl TE for 30 min at 37 °C or overnight at 4 °C.

9 Remove 1 μl of resuspended DNA pellet for the assessment of the labelling quality (*Protocol 12*). Precipitate and store the rest of the reaction as follows (steps 10–13).

10 Add 551 μl TE, 5 μl SS-DNA (10 mg/ml), 5 μl tRNA (10 mg/ml), and 61 μl of 3 M NaAc pH 5.5. Divide the solution into two tubes (335 μl each), add 840 μl of 100% EtOH (–20 °C) to each tube, and incubate for 30 min on ice.

11 Centrifuge for 30 min at 3750 g and 4 °C.

12 Dry pellet for 5 min in Speed Vac and then dissolve in 50 μl FSP.

13 Store at –20 °C.

14 Assess the quality of the labelling using a spot test (*Protocol 12*).

[a] This protocol refers to the labelling of the probes with digoxigenin or biotin, but also applies to other labelling methods (i.e. direct labelling protocols).

[b] Add reaction components on ice.

Protocol 12

Determination of labelling quality (spot test)

Reagents

• See *Protocol 11*

• Hybond-N+ (Amersham, RPN 203B)

Method

1 Use 1 μl of the resuspended probe (*Protocol 11*, step 9) for probe labelling examination. Prepare a dilution series of the probe and compare it with a standard (i.e. the identical dilutions of a control probe or the DIG labelled control DNA from Kit).

2 Spot 1 μl of each dilution onto a dry piece of Hybond-N+.

3 Dry the membrane and bake it for 2 h at 80 °C under vacuum.

4 Wash the membrane for 1 min in buffer 1, with shaking.

5 Block membrane for 30 min in buffer 2, with shaking.

6 Incubate membrane for 30 min in anti-DIG–AP antibody (or the appropriate primary antibody) with shaking. (The working dilution of DIG–AP antibody is 1/5000 in buffer 2, for streptavidin–AP antibody, 1/25 000 in buffer 2.)

7 Wash membrane twice for 15 min in buffer 1.

8 Equilibrate membrane for 2 min in buffer 3.

9 Place 5 ml substrate solution on the membrane.

10 Incubate the membrane in the dark until spots are clearly visible.[a]

11 The newly labelled probe is titrated against a former probe of the same labelling quality in a FISH hybridization (see *Protocol 13*).[b]

[a] The reaction is stopped after 5 min with $1 \times$ TE buffer, the membrane air dried, and kept for records.

[b] In general, equally labelled identical hybridization probes show similar hybridization qualities; i.e. if 20 ng of a probe gave a good hybridization signal, the new probe will lie in a similar range (i.e. around 20 ng). However, such titration experiments following each labelling are done to ensure the optimal concentration of a specific probe. Optimal refers to specific hybridization signals, hybridization efficiencies, and absence of background hybridization.

Protocol 13

Fluorescent *in situ* hybridization

Reagents

- HCl (Fisher, A1445-225)
- RNase A (DNase-free) (Sigma, R6513)
- Pepsin (Sigma, P7012), 50 μg/ml
- Formaldehyde (Fluka, 47629)
- Formamide (Fluka, 47670)
- BSA (Fluka, 05480)
- Mouse Cot-1 DNA (Invitrogen, Y01398)
- Resin (Bio-Rad, 143-6425)
- Lamb serum (Invitrogen, 160 70-096)
- Tween 20 (Sigma, P1379)
- $2 \times$ SSC: 0.3 M NaCl, 0.03 M Na$_3$.citrate pH 7.6
- Phosphate-buffered saline (PBS): 0.15 M NaCl, 1.4 mM NaH$_2$PO$_4$, 4.3 mM Na$_2$HPO$_4$, 2.7 mM KCl pH 7.3
- Hybridization solution: see *Protocol 11*[d]
- 70%, 90%, 100% EtOH

Method

1 Fix slides in 50% acetic acid 6×20 min at room temperature.

2 Equilibrate slides in $2 \times$ SSC for 10 min at room temperature.

3 Incubate slides with 100 μg/ml RNase A for 1 h at 37 °C in a humidified incubator.

4 Wash slides three times in $2 \times$ SSC for 5 min with shaking at room temperature.

5 Incubate slides in fresh 0.01 M HCl and pepsin (50 μg/ml) for 10 min at 37 °C.[a,b]

6 Wash slides twice for 5 min in PBS, while shaking at room temperature.

Protocol 13 continued

7   Wash slides in PBS, 50 mM $MgCl_2$ for 5 min at room temperature.

8   Incubate slides for post-fixation in 1% formaldehyde in PBS, 50 mM $MgCl_2$ at room temperature for 10 min.

9   Dehydrate slides for 3 min cycles in 70%, 90%, and 100% EtOH at room temperature and air dry.

10 Denature slides in 70% deionized formamide/2 × SSC at 70 °C, pH 7.0[a,c] for 2 min.

11 Dehydrate slides for 3 min cycles in 70%, 90%, 100% cold (–20 °C) EtOH at room temperature.

12 Denature probe in hybridization solution (20 μl)[d,e] and add to slide.

13 Hybridize overnight at 37 °C in a humidified atmosphere.

14 Wash slides at 42 °C in 50% formamide, 2 × SSC.

15 Repeat twice.

16 Wash slides in 2 × SSC for 2 min at room temperature, with shaking.

17 Repeat four times.

18 Block slides in 4 × SSC, 4% BSA for 10 min at room temperature.[f]

19 Add antibodies for detection of hybridization signals as outlined in *Protocol 14*.

[a] This solution is always freshly prepared.

[b] The pepsin concentration may vary, depending on cell type. One attempts a balance between morphology and hybridization efficiency.

[c] It is crucial to work with deionized formamide. In our laboratory, we deionize the formamide (Fluka) for 6 h with 50 g resin (Bio-Rad) /litre formamide, filter it and store it in aliquots at –20 °C until further use.

[d] The pH is critical for good denaturation and subsequent hybridization results.

[e] Probes containing repetitive sequence motifs require the presence of Cot-1 DNA during denaturation and hybridization. We add 50 ng mouse Cot-1 DNA/20 ng probe/slide and pre-anneal the probe mixture following denaturation for 2 h at 37 °C before adding to the target.

[f] The following additional blocking reagents have been used successfully in our laboratory: (a) lamb serum, it is relatively inexpensive and very efficient in blocking reactions, especially if one encounters some background problems; (b) 4%BSA, 4 × SSC, 0.1% Tween 20. Note that blocking with Tween 20 is more stringent than the same reagent without Tween 20.

Protocol 14

Detection of hybridization following FISH (see *Figure 2*)

Reagents

- DAPI (Sigma, D9542)
- Anti-DIG–FITC antibody (Roche, 1207-741)
- Anti-biotin antibody (Roche, 1297-597)
- Anti-mouse IgG1–Texas Red antibody (Southern Biotechnology Assoc. Inc., 1030-07)

Protocol 14 continued

Method

1 Block non-specific antibody binding by incubating slide in 4% BSA, 4 × SSC for 10 min at room temperature.

2 Titrate each new antibody used for the detection of labelled probe against a previous batch of the same antibody.[a]

3 Add 100 μl of the diluted antibody to the hybridization area of the slide and cover with a coverslip (24 × 60 mm).

4 Incubate for 30 min at 37 °C in a humidified atmosphere.

5 Wash off the unbound antibody three times in 4 × SSC, 0.1% Tween 20 for 5 min at 42 °C.[b]

6 Counterstain with DAPI and mount in antibleach/antifade.

[a] Commonly used antibodies are anti-DIG–FITC at 1:200 or 1:250 dilutions; anti-biotin at 1:100 or 1:200; anti-mouse–TXRD 1:400 dilutions.

[b] If additional antibodies are required to amplify the signal, block the slides as described in step 1 and proceed as described (*Protocol 14*, steps 2–6).

Protocol 15

Chromosome painting (see *Figure 3*)

Reagents

• Chromosome paints (CedarLane)

Method

1 Perform *Protocol 13*, steps 1–11.

2 Add 50 ng denatured mouse Cot-1 per probe reaction prior to denaturation if necessary.[a]

3 Denature the chromosome paint for 8 min at 85 °C.

4 Transfer paints for pre-annealing to 37 °C for 30 min.

5 Perform *Protocol 13*, steps 13–19.

6 Detect hybridization of paint(s) as described in *Protocol 14*.

[a] Some suppliers have Cot-1 DNA included in the paint containing vial and the addition of the Cot-1 DNA to the paint is therefore not required.

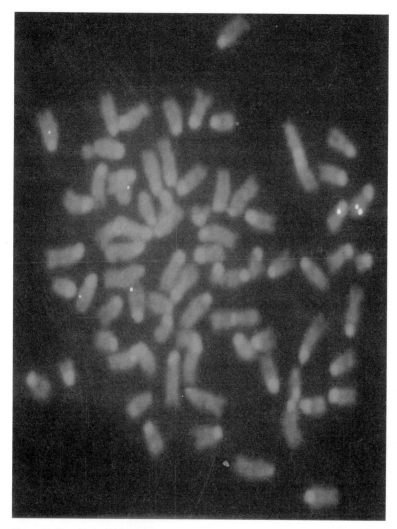

Figure 2 (See *Plate 4*) Mouse metaphase chromosomes hybridized with *DHFR* (19) that was labelled by digoxigenin and visualized by an anti-digoxigenin–FITC antibody. For details, see *Protocols 13* and *14*.

Protocol 16

Spectral karyotyping (SKY) (see *Figure 4*)

Reagents

- Mouse SKY kit as sold by Applied Spectral Imaging (ASI, SKY-M10)

Method

1 Perform *Protocol 13*, steps 3–11.

2 Follow the instructions of hybridization and detection of hybridization signals as specified by the manufacturer.

3 Follow the banding procedure described in *Protocol 17*.

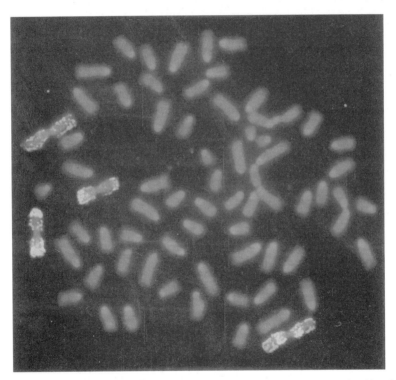

Figure 3 (See *Plate 5*) Painting of primary plasmacytoma metaphase spreads with paints for chromosomes 6 (red) and 15 (green). Note the reciprocal translocation (rcpT Rb6;15). For details, see *Protocol 15*.

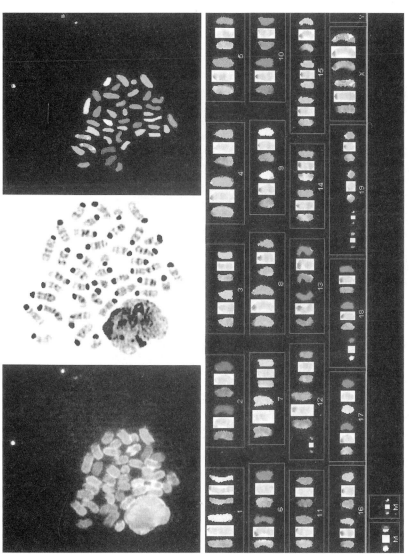

Figure 4 (See *Plate 6*) Banding of mouse chromosomes after SKY. For details see *Protocols 16* and *17* and ref. 21.

Protocol 17

Banding after SKY (21)

Reagents

- Antibleach: 12% glycerol (Fisher, G-31), 4.8% Mowiol 4-88 (Calbiochem, 475904), 2.4% DABCO (1,4 diazabicyclo[2.2.]-octane) (Fluka, 33480) in 0.2 M Tris (Sigma, T-1503) HCl pH 8.5

- DAPI working solution: 1 μg/ml in 1 × PBS
- Quinacrine mustard (Sigma, Q2876) working solution: 1 μg/ml in PBS pH 7.5.

Method

1 Immerse the SKY painted slide into the quinacrine mustard staining solution for 10 min at room temperature.

2 Wash slide twice in PBS at room temperature.

3 Counterstain slide with DAPI for 5 min at room temperature.

4 Wash slides once in PBS at room temperature.

5 Mount slide in antibleach.

References

1. Sandberg, A. A. (1980). In *The chromosomes in human cancer and leukemia* (1st edn) (ed. B. A. Conover), p. 98. Elsevier North Holland Inc.

2. Lee, J. J., Warburton, D., and Robertson, E. J. (1990). *Anal. Biochem.*, **5**, 1.

3. Committee on Standard Genetic Nomenclature for Mice. (1979). *Mouse News Lett.*, **61**, 4.

4. Jotterand Bellomo, M., Muhlematter, D., and Nabholz, M. (1990). *Genetica*, **83**, 51.

5. Fukasawa, K., Wiener, F., Vande Woude, G. F., and Mai, S. (1997). *Oncogene*, **15**, 1295.

6. Spurbeck, J. L., Zinsmeister, A. R., Meyer, K. J., and Jalal, S. M. (1996). *Am. J. Med. Genet.*, **2**, 387.

7. Hliscs, R., Muhlig, P., and Claussen, U. (1997). *Cytogenet. Cell Genet.*, **76**, 167.

8. Potter, M. and Wiener, F. (1992). *Carcinogenesis*, **13**, 1681.

9. Lee, Y. F. and Arrighi, P. E. (1981). *Chromosoma*, **83**, 721.

10. Sen, P. and Sharma, T. (1985). *Cytogenet. Cell Genet.*, **39**, 145.

11. Drouin, R., Lemieux, N., and Richer, C. L. (1991). *Cytogenet. Cell Genet.*, **57**, 91.

12. Koshland, D. and Strunnikov, A. (1996). *Annu. Rev. Cell Dev. Biol.*, **12**, 305.

13. Sawyer, J. R., Moore, M. M., and Hozier, J. C. (1987). *Chromosoma*, **95**, 350.

14. Cowell, J. K. (1984). *Chromosoma*, **89**, 294.

15. Comings, D. E. and Drets, M. E. (1976). *Chromosoma*, **56**, 199.

16. Sumner, A. T. (1980). *J. Microsc.*, **119**, 397.

17. Fukushima, Y. (1986). *Hakkaido Igaku Zasshi*, **61**, 935.

18. Chang, A. C., Nunberg, J. H., Kaufman, R. J., Erlich, H. A., Schimke, R. T., and Cohen, S. N. (1978). *Nature*, **275**, 617.

19. Mai, S. (1994). *Gene*, **148**, 253.

20. Mai, S., Hanley-Hyde, J., and Fluri, M. (1996). *Oncogene*, **12**, 277.

21. Wiener, F. and Mai, S. (2000). Technical Tips Online. http://www.biomednet.com/db/tto. T01884.

Chapter 5

High resolution FISH mapping using chromatin and DNA fibre

Henry H. Q. Heng

Center for Molecular Medicine & Genetics, Wayne State University School of Medicine, 5107 Biological Science Building, 5047 Gullen Mall, Detroit, MI 48202, USA.

1 Introduction

This chapter focuses on the method of high resolution fibre FISH. Fibre FISH collectively refers to all approaches that perform fluorescence *in situ* hybridization on released chromatin or DNA fibre. Since the degree of condensation is much less than metaphase chromosomes and even less than interphase chromatin, the resolution of fibre FISH is among the highest when compared to prophase chromosome FISH, interphase FISH, meiotic prophase FISH, or pronuclei FISH (1–5).

As a new experimental approach to visualize the detail of centromeric sequences on elongated chromatin, fibre FISH was first introduced into the scientific community in 1991 called 'free chromatin FISH' (6). Various modifications immediately followed to improve resolution and to simplify the procedures. Major improvements developed with the release of chromatin and DNA fibres by simply lysing the nuclei or by extracting protein for DNA 'halo' preparations or from lysing nuclei in gel blocks and finally by producing free DNA molecules within solution that are ready for application to slides (7–15). Consequently, different names were used for the description of these modified procedures including free chromatin FISH, free DNA FISH, DNA halo FISH, extended chromatin FISH, extended DNA fibre FISH, elongated chromatin FISH, elongated DNA FISH, individual stretch DNA molecule FISH, visual mapping, direct visual hybridization (DIRVSH), quantitative DNA fibre mapping, and dynamic molecular combing. The original and alternative approaches are now collectively referred to as fibre FISH (16, 17) since the key difference between them is the method of chromosome/DNA fibre preparation and the corresponding resolution.

FISH represents a cutting edge experimental system for direct visualization of specific DNA sequences in either the chromatin/chromosome or in the nucleus (18). Since its introduction high resolution fibre FISH has been extensively used for physical mapping of both plant and animal genomes (1, 19, 20). In particular, fibre FISH has been used for the establishment of contigs, mapping ESTs, filling gaps in sequence ready maps, and estimating the size of the gaps remaining in sequences for the current Human Genome Project (21–23).

Fibre FISH has powerful applications in chromosomal structure studies (1), where high resolution is a key feature filling the gap between molecular biology and cytology. These research projects include the analysis of the centromere, telomere, measuring the size of transgenic insertions (24, 25), the study of the integration patterns of the virus in host genomes (26), and visualizing the DNA–protein interaction by DNA–protein co-visualization (27). Fibre FISH has also been used in medical genetics for the study of gene amplification, deletion, and translocation, as well as the correlation of gene numbers with clinical phenotypes in diseases caused by additional copy numbers of genes (28, 29).

2 Practical considerations for fibre preparation

Determining the type of fibre (chromatin or DNA) to be generated and utilized as a substrate is the first consideration. For defined physical mapping that ranges from a few kb to a few hundred kb, DNA fibre should be selected since DNA fibre provides the highest resolution but covers the smallest area. In contrast, when mapping a large area in the range of 50 kb to greater than a few hundred kb, chromatin fibre would be the substrate of choice. This is due to the fact that it is very difficult to trace DNA fibre when the target size is large. A further consideration is that chromatin fibre is more frequently used in chromatin structure studies to evaluate higher levels of organization than when performing physical mapping.

After selecting the type of fibre, the means of releasing the fibre is considered next. There are numerous protocols utilized by various laboratories to accomplish this. These protocols all come under the following general categories:

(a) Releasing fibre by chemical treatment either by lysing the nuclear membrane or releasing DNA fibre by extracting protein.

(b) Physical stretching.

(c) A combination of (a) and (b).

Protocols have been developed that also vary the timing of fixation in terms of before or after the release of fibre. In addition, protocols also differ by the release of fibre in solution, in gel blocks, or directly on the slide. Some of these strategies will be used in the following protocols.

3 General equipment required for fibre FISH

Some basic equipment is needed, this includes:

(a) Epifluorescence microscope.
   i. Leitz Aristoplan epifluorescence microscope with DAPI filter, dual bandpass FITC/Texas Red filter (Omega Optical Inc.), or tri-pass DAPI/ FITC/ rhodamine filter (optional).
   ii. CCD and image system (optional).

(b) Other essential items.
   i.   Plastic jars (25 ml slide mailers, Surgipath).
   ii.   Incubator (37 °C setting).
   iii. Three water-baths (37, 43, 70–75 °C settings).
   iv.   Frosted slides (Fisher Scientific).
   v.   Coverslips (Fisher Scientific).
   vi.   A small light-proof box for slide incubation.
   vii. A phase-contrast microscope.
   viii. Cytospin (Shandoz).

4 Chromatin fibre preparation

Many protocols may be used to prepare chromatin fibre from cells (1, 30, 31). The release of chromatin fibre can be summarized into four different categories. These are the following:

(a) Interfering with chromosomal condensation by chemical treatments.

(b) Lysing the nuclear envelope by utilizing alkaline buffer or extensive hypotonic treatment.

(c) Selecting the late G2 or early G1 phase to accumulate free chromatin.

(d) Physically stretching the nuclei during slide production by using a cell centrifuge.

Three protocols have been selected to represent these studies and will be detailed. The drug treatment approach is the original method that increased the frequency of free chromatin that was originally observed (32, 33). The experimental steps for this approach are tedious when compared to other methods, however different drugs may be used to delineate the chromosome condensation process. The high pH buffer approach is recommended for obtaining chromatin fibre from cultured fibroblast cell lines which are less sensitive to drug treatment when compared to lymphocytes. Using cytospin to generate chromatin fibre works well for all cell lines.

To increase the number of chromatin fibres, improved protocols have been developed to open the nuclear envelope. The rationale for the use of alkaline buffer to release chromatin fibres from cultured cells is based on the observation that nuclear lamins can be disrupted by high pH treatment (7). It is thought that the treatment destabilizes the nuclear envelope, which then can be easily ruptured by the hypotonic solution. Prolonged alkaline treatment, however, may break the nuclear envelope even without hypotonic shock. Although the alkaline protocol was originally designed for fibroblastoid cell cultures, it can also be used for other cell types including lymphocytes.

Protocol 1

Preparation of chromatin fibre from cultured lymphocytes by drug treatment

Equipment and reagents

- CO_2 incubator
- Water-bath (37°C)
- T25 culture flask
- Slides (Fisher Scientific)
- Clinical centrifuge
- 15 ml centrifuge tube
- 0.4% KCl
- RPMI 1640 (Invitrogen)
- FCS (fetal calf serum) (Invitrogen)

- PHA (phytohaemagglutinin) (Invitrogen)
- Penicillin/streptomycin (Invitrogen)
- BrdU (5-bromodeoxyuridine) (Sigma)
- EtBr (ethidium bromide) (Sigma)
- mAMSA (N-[4-(9-acridinylamino)-3-methoxyphenyl] methanesulfonamide) (NIH)
- Fixative: methanol/acetic acid, 3:1 (Fisher Scientific)

Method

1 Isolate lymphocytes from 5–10 ml of fresh human peripheral blood by a low speed centrifugation (50 g for 5 min) or by unit-gravity sedimentation. Then collect the white cell suspension.[a]

2 Transfer 0.2–0.4 ml of the white cell suspension to 10 ml of RPMI 1640 in a T25 culture flask containing 10% FCS, 100 U penicillin/streptomycin, and 1% PHA.

3 Culture the lymphocytes in a CO_2 incubator at 37°C for 48–52 h.

4 Add mAMSA, or EtBr, or BrdU (10 μg/ml) to cultured lymphocytes for 2–4 h.[b]

5 Transfer the culture to two 15 ml centrifuge tubes (5 ml each) then collect the cells by centrifugation at 1000 g for 7 min.

6 Remove the supernatant down to 0.3 ml then mix the pellet.

7 Add 5 ml of 0.4% KCl solution, mix, and incubate in a 37°C water-bath for 15 min.

8 At the end of the hypotonic treatment, add 0.1–0.2 ml of freshly prepared fixation solution (methanol/acetic acid, 3:1) for pre-fixation at room temperature. Mix the contents gently by inverting the tube, and then pellet the cells again by centrifugation (1000 g for 7 min).

9 Discard the supernatant. Loosen the cell pellet by gently tapping the bottom of the tube.

10 Resuspend the cells in 5 ml of the fixation solution. Fix at room temperature for 20 min.

11 Spin down the cells, discard the supernatant, and resuspend the pellet in 0.5 ml of fresh fixation solution.

12 Drop the fixed cell suspension on to the surface of a pre-chilled microscope slide (placed on ice for 10 min).

Protocol 1 continued

13 Air dry the slide.

14 Check the density of the chromatin fibre on the slide. Adjust the concentration of the suspension by adding more fixation solution if the chromatin fibres are too crowded on the slide.

15 Make additional slides.[c]

[a] 0.5 ml of heparinized whole blood can be used without isolating white cells. We recommend using human cord blood.

[b] mAMSA, BrdU, or EtBr, can be replaced with each other. mAMSA can be obtained from the Drug Synthesis Branch, National Cancer Institute, Bethesda, MD, USA. Prepare stock solution by dissolving mAMSA in dimethyl sulfoxide at 10 mg/ml, diluting with an equal volume of distilled water, and filter sterilizing. The concentration we recommend for these drugs (mAMSA, or BrdU, or EB) is 10 μg/ml. If needed, the concentration can vary between 5–15 μg/ml for different individuals.

[c] Slides containing free chromatin fibre preparations may be stored for several weeks at -20 °C. Once a good batch of slides is obtained, they should be dried at room temperature for one day and then sealed in slide containers with Parafilm before transferring them to the freezer. Over-dehydrated slides tend to irreversibly damage the chromatin fibres.

Protocol 2

Preparation of chromatin fibres with alkaline buffer

Equipment and reagents

- Centrifuge
- CO_2 incubator
- Microscope slides (Fisher Scientific)
- Water-bath
- Alkaline lysis buffer: 1 mM sodium borate solution adjusted to pH 10–11 with NaOH, mixed with KCl solution (0.4%) in a 1 : 1 ratio before use

- 0.4% KCl (Fisher Scientific)
- Methanol (Fisher Scientific)
- Acetic acid (Fisher Scientific)
- α-MEM (Invitrogen)
- Fetal calf serum (FCS) (Invitrogen)
- Trypsin/EDTA (Invitrogen)

Method

1 Grow fibroblasts in α-MEM with 5% FCS for two to four days after reaching confluence in order to accumulate cells arrested at the G1 phase.

2 Add 1 ml of the trypsin/EDTA (1 ×) solution to each 60 mm Petri dish then incubate at 37 °C for 30–60 sec.

3 Add 1 ml of culture medium with serum once the majority of the cells become detached from the dish. Proceed quickly to the next step.

4 Prepare a series of tubes containing 1 ml of alkaline buffer each.[a]

Protocol 2 continued

5 Transfer an aliquot (0.2–0.3 ml) of the trypsinized cell suspension into each of these tubes in drops, tap the tube gently to mix the contents.

6 At various time intervals (3–5 min), the alkaline treatment is terminated by adding 5 ml of fixation solution (methanol/acetic acid, 3:1).

7 Collect the alkaline treated cell suspension by centrifugation (1000 g for 7 min).

8 Resuspend the cells in 5 ml of new fixation solution.

9 Spin down and discard the supernatant. Resuspend the pellet in 0.1–0.2 ml of fixation solution.

10 Drop the suspension on to the pre-chilled slide (placed on ice for 10 min).

11 Air dry the slide and check the density of the chromatin fibre on the slide using a phase-contrast microscope.

12 Adjust the concentration of the suspension by adding more fixation solution if needed.

13 Make slides. The slides can be used for FISH immediately or can be stored at –20 °C for later usage.

[a] For alkaline treatment the optimal condition for each particular cell line has to be obtained empirically and a brief screening test is recommended. Many factors affect the generation of chromatin fibres including the length of treatment of the alkaline buffer and the pH of the buffer (10–11.5). In general, a high pH and longer treatment time promotes the lysis of nuclei. The conditions can be varied systematically to find the optimal combination for a given cell line. Over-treatment should be avoided since it reduces the number of useful chromatin fibres. Over-treatment will also destroy the 300 Å structure and lead to the production of naked DNA fibres. To obtain optimal results, one should attempt to fix the chromatin fibres quickly before they become aggregated with each other after release from the nuclei. The use of a small volume of alkaline buffer and a large volume of the pre-fixation solution is a good way to avoid aggregation of chromatin fibres.

Protocol 3

Preparation of chromatin fibre using cytospin

Equipment and reagents

- Microscope slides (Fisher Scientific)
- Centrifuge
- CKS buffer: 10 mM Pipes pH 7–8, 100 mM NaCl, 0.3 M sucrose, 3 mM $MgCl_2$, 0.5% Triton X-100
- Cytospin
- Ice
- Fixative: methanol/acetic acid, 3:1 (Fisher Scientific)

Method

1 Resuspend the cells to $1–5 \times 10^6$ cells in 1.0 ml CSK buffer and incubate on ice for 15 min.[a]

2 Load microscope slides into the clips and place a cytofunnel with filter paper against the slide. Secure the metal clip and place into a rotor.

Protocol 3 continued

3 Load 100 μl of cells in the CSK buffer into the cytofunnel.

4 Cytospin the cells on to microscope slides at 1000 g for 7 min.

5 Rinse the slides with fixation solution (methanol/acetic acid, 3:1).

6 Check the slide using the phase-contrast microscope. Good slides can be used for FISH immediately or stored at −20 °C.[b]

[a] For monolayer cultures, use trypsin/EDTA treatment to harvest cells and wash cells twice with PBS. For suspension cultures, harvest cells by centrifuge and wash cells twice with PBS. Cells in PBS can be stored at 4 °C for hours before use.

[b] The ideal cell concentration for slide making varies for different types of cells. The concentration can be adjusted accordingly. If it is too crowded, reduce the cell number in CSK buffer. If there are too few cells then add more cells in CSK solution or load more cells into the cytofunnel.

5 DNA fibre preparation

Many modified protocols have been published to prepare released DNA fibre. These protocols can be classified into the following categories:

(a) Complete releasing of DNA using detergent/alkaline treatment following linearized fibre produced by gravity or pulling with a coverslip.

(b) Generating DNA fibre from a 'halo' preparation by protein extraction.

(c) Lysing nuclei and releasing DNA in a gel block and linearizing fibres by a mechanical or an electronic pulling force.

(d) Linearize DNA molecules by 'water–air phase interaction'.

The two protocols outlined below are currently very popular due to their simplicity and the quality of fibres produced. *Protocol 4* is one of the simplest protocols to implement (10). *Protocol 5* places lysed nuclei in a gel block in order to reduce the interference from other nuclei (14). These two protocols should be useful encompassing most cases and can serve as a technical gate for additional protocols.

Protocol 4

DNA fibre preparation

Equipment and reagents

- Centrifuge
- Microscope slides (Fisher Scientific)
- 0.4% KCl (Fisher Scientific)
- Fixative: methanol/acetic acid, 3:1 (Fisher Scientific)
- Alkaline solution: 0.07 M NaOH/ethanol (5:2)
- PBS solution (Invitrogen)
- Methanol (Fisher Scientific)
- 70%, 95%, and 100% ethanol

Protocol 4 continued

Method

1 Transfer the cultured cells into a 15 ml centrifuge tube. Collect the cells by centrifuging at 1000 g for 7 min.

2 Resuspend the cell pellet in 0.3 ml of medium, and add 4 ml of 0.4% KCl for 10 min at 37°C.

3 Collect the cells at the end of the hypotonic treatment by centrifugation.

4 Discard the supernatant.

5 Add 5 ml of fixative and mix well.

6 Fix once more in fresh fixative (20 min).

7 Drop fixed cells on to a slide.

8 Soak the slide in 1 × PBS for 1 min at room temperature (do not let the slide dry out).

9 Add 200 μl of alkaline solution to the slide.

10 Place one edge of the coverslip on to the slide and pull it along the slide from one end to the other.

11 Add one drop of methanol to the slide.

12 Soak the slide in a 70%, 95%, 100% series of ethanol.

13 Air dry. The slides can then be used for FISH.

Protocol 5

DNA fibre preparation using gel blocks

Equipment and reagents

- Microwave oven
- Poly-L-lysine-coated slide (Sigma Chemical)
- Low melting point agarose (FMC BioProducts)
- Proteinase K (Merck)
- 50 mM EDTA
- N-lauroylsarcosine (Sigma Chemical)
- TE buffer: 1 mM EDTA, 10 mM Tris–HCl pH 7.5
- RNase A (Boehringer Mannheim)
- Standard saline citrate (SSC)

Method

1 Embed lymphocytes in 1% low melting point agarose for the preparation of blocks containing 1×10^7 cells/ml.

2 Incubate the blocks of cells at 50°C in 1 mg/ml proteinase K in 50 mM EDTA in the presence of 1% N-lauroylsarcosine for 48 h.

3 Wash the blocks in several changes of 1 × TE buffer over two days.

Protocol 5 continued

4 Treat with 100 μg/ml RNase A in 2 × SSC at 50 °C overnight.

5 Store at 4 °C in 50 mM EDTA.

6 Place a small piece of agarose-embedded DNA (1/8 of the 100 μl block containing about 5 μg DNA) at the end of a poly-L-lysine-coated slide.

7 15 μl of water is added to the block.

8 Heat the slide in a microwave oven until the agarose melts (about 30 sec).

9 Extend the DNA on the slide with the aid of another slide.

10 Air dry.

6 FISH

Fibre FISH can be broken down into two steps including fibre preparation and probe hybridization/detection. The probe hybridization and detection protocols are essentially the same for metaphase and interphase FISH.

6.1 DNA probe labelling

Protocol 6
Probe labelling

Equipment and reagents

- Centrifuge
- Digoxigenin (DIG)–nick translation mix (Roche Diagnostics)
- Biotin labelling nick translation kit (Invitrogen)
- Nick column (Pharmacia)
- Equilibration buffer: 10 mM Tris–HCl pH 7.5, 1 mM EDTA
- Salmon sperm DNA (10 μg/μl) (Invitrogen)
- 3 M sodium acetate (NaAc)
- 70% and 100% ethanol
- TE buffer pH 7.5

Method

1 Label DNA using BRL BioNick kit or Boehringer DIG–nick translation mix according to the supplier's instructions.[a]

2 Perform the labelling reaction at 15 °C for 1 h for smaller probes or 2 h for large probes (BAC/PAC/YAC).

3 Stop the reaction by adding stop buffer.[b]

4 Separate the unincorporated nucleotides from the labelled probe by using a Nick column. Load the labelled products on an equilibrated column (with equilibration buffer). Add 400 μl equilibration buffer and let it enter into the gel bed. Elute the purified sample with 400 μl equilibration buffer.[c]

Protocol 6 continued

5   Add 6 μl of sonicated salmon sperm DNA (60 μg) to the eluted probe solution along with 40 μl of 3 M NaAc.

6   Precipitate with the addition of 880 μl cold ethanol.

7   Wash the pellet with 70% ethanol then air dry.

8   Resuspend the pellet in 20 μl of 10 mM TE buffer pH 7.5.

[a] Biotin and digoxigenin nick translation labelling kit can be prepared in-house as follows: DNA polymerase I, DNase I, 10 × dNTP mix (0.2 mM dATP, 0.2 mM dCTP, 0.1 mM dGTP, and 0.1 mM dTTP), 10 × reaction buffer (500 mM Tris–HCl pH 7.8, 50 mM MgCl$_2$, 100 mM 2-mercaptoethanol, 100 μl/ml bovine serum albumin), and biotin-16-dUTP or DIG-11-dUTP. For labelling 1 μg of probe DNA, mix 5 μl of 10 × reaction buffer, 2.5 μl of biotin-16-dUTP, 5 μl 10 × dNTP mix 20 U of DNA polymerase I, and diluted DNase I (1:1000). Bring the total reaction volume to 50 μl with dH$_2$O. Enzyme solution is the last ingredient to be added. For labelling probe with digoxigenin, use 5 μl of DIG-11-dUTP instead of biotin-16-dUTP.

[b] The size of the products of nick translation can be checked with 1% agarose mini gel. The optimal size of labelled fragments is approx. 150–250 bp.

[c] The separation of incorporated nucleotides using Nick column is an optional step.

6.2 Hybridization

Protocol 7

Probe denaturation and pre-hybridization

Equipment and reagents

- Water-baths at 37 °C, 43 °C, 70 °C, and 75 °C
- Human Cot-1 DNA (Invitrogen)
- Super Blocker DNA (SeeDNA Biotech Inc.)
- Hybridization solution I (for use with genomic probes): 50% deionized formamide (Fluka) and 10% dextran sulfate in 2 × SSC

- Denaturation solution: 70% deionized formamide (Fluka) in 2 × SSC (saline sodium citrate)
- 20 × SSC stock solution: 3 M NaCl, 300 mM Na citrate
- Hybridization solution II (for use with repetitive DNA probes): 65% formamide (Fluka) and 10% dextran sulfate in 2 × SSC

A. For non-repetitive DNA probes

1   Mix 20–50 ng of plasmid or cosmid or phage DNA or 200–250 ng of BAC/PAC/YAC DNA with 2 μg of Cot-1 DNA[a] or Super Blocker[b] in 30 μl of hybridization buffer I.

2   Denature at 75 °C for 5 min.

3   Incubate the mixture at 37 °C for 15–30 min for probe pre-hybridization.

Protocol 7 continued

B. For repetitive DNA probes

1 Mix 20–50 ng of probe DNA with 30 μl of hybridization buffer II.

2 Heat the mixture at 75 °C for 5 min and transfer to ice immediately before use.

[a] For probes containing dispersed repetitive DNA sequences (such as Alu or L1), suppression by pre-hybridization to total genomic DNA or Cot-1 DNA is required. Pre-hybridization is not necessary for repetitive probes such as α-satellite sequences.

[b] 'Super Blocker' is a cocktail of repetitive DNA developed to compete with and block out signals from repetitive elements during FISH analysis and Southern blotting. This effective blocker reagent can be used in place of Cot-1 DNA. Information can be obtained from SeeDNA Biotech Inc.

Protocol 8

Slide preparation for hybridization

Equipment and reagents

- Slide jar
- Water-bath
- RNase A: 100 μg/ml in 2 × SSC in a 25 ml jar (Invitrogen)
- Ethanol: 75%, 90%, and 100%

- Denaturation solution: 70% deionized formamide (Fluka) in 2 × SSC
- 20 × SSC stock solution: 3 M NaCl, 300 mM Na citrate

Method

1 Bake slides of the chromatin or DNA fibre at 55 °C for 1–2 h just before use.

2 Treat the baked slides with RNase A at 37 °C for 1 h.

3 Dehydrate by dipping them in 75%, 90%, and 100% ethanol consecutively for 3 min each.

4 Air dry.

5 Immerse them for 2–4 min in freshly prepared denaturation solution pre-heated to 72 °C (use a 25 ml slide jar secured in a water-bath).[a,b]

6 Quickly transfer the slides into a jar containing ice-cold 70% ethanol for 2 min.

7 Dehydrate the slides by going through 95% and 100% ethanol for 3 min.

8 Air dry.

9 Proceed with the hybridization immediately.

[a] Several slides may be treated at the same time. When denaturing additional slides a slightly higher denaturation temperature is needed since each slide may lower the temperature about 1 °C.

[b] For FISH detection on metaphase chromosomes, it is important to avoid over-denaturation by limiting the denaturation time (34, 35). For the denaturation of fibre slides a longer time interval is suggested.

Protocol 9

Hybridization

Equipment and reagents

- Coverslip 22 × 22 mm
- 37 °C moist chamber
- Rubber cement
- Hybridization solution

Method

1 Load the hybridization solution with probe to each slide.

2 Cover the hybridization area with a 22 × 22 mm coverslip, avoid air bubbles, and seal the edges with rubber cement to prevent evaporation.

3 Incubate the slides at 37 °C in a moist chamber containing absorbent paper soaked in 2 × SSC.

4 For repetitive DNA probes, hybridization is allowed for 3–16 h.

5 For non-repetitive DNA probes, the incubation time should be 12–24 h.

6.3 Wash

Prepare three plastic jars filled with washing solution and three jars of 2 × SSC pre-warmed to 43 °C. Different conditions are used to wash the slides after hybridization, according to the nature of the probe.

Protocol 10

Wash

Equipment and reagents

- Water-bath
- Plastic jars
- 2 × SSC (saline sodium citrate)
- Wash solution A (for non-repetitive DNA probes): 50% formamide in 2 × SSC
- Wash solution B (for repetitive DNA clones): 65% formamide in 2 × SSC
- Wash solution C: 0.1 M phosphate buffer pH 8.0, with 0.1% Nonidet P-40 (Roche Diagnostics)

A. Non-repetitive DNA probes

1 Carefully remove the rubber cement from the slides by forceps.

2 Dip the slide in 2 × SSC and allow the coverslips to float off.

3 Agitate the slides in the solution a few times.

4 Immerse the slides in the pre-warmed (43 °C) washing solution A, three times for 3 min each (gently agitate).

5 Wash in 2 × SSC for three times at 43 °C for 3 min each.

6   Place slides in wash solution C at room temperature. The slides are ready for detection.

B. Repetitive DNA probes

1   Pre-warm washing solution B and 2 $\times$ SSC to 43 °C.

2   Follow part A, steps 1–6. In step 4, replace solution A with solution B.

6.4 Detection and amplification

Protocol 11

Detection with signal amplification[a]

Equipment and reagents

- Incubator
- Plastic jars
- Blocking solution: 3% BSA (bovine serum albumin, Sigma Fraction V)
- FITC–avidin (fluorescein isothiocyanate, Vector): 500 μg/ml (stock solution). FITC detection working solution: 10 μl of avidin–FITC stock solution to 990 μl detection mixture. Store in the dark at 4 °C. Good for up to six months.
- Wash solution C: 0.1 M phosphate buffer pH 8.0, with 0.1% Nonidet P-40 (Roche Diagnostics)

- Biotinylated goat anti-avidin antibody (Vector): 500 μg/ml (stock solution). Aliquots (50 μl each) can be kept at –20 °C. Working solution: dilute stock solution with detection solution. The final concentration of the solution with detection mixture is 5 μg/ml.
- Detection solution: 1% BSA and 0.1% Tween 20 in 4 $\times$ SSC; store at 4 °C

Method

1   Take out the slides from washing solution C and blot excess liquid from the edges of the slides.

2   Apply 60 μl of blocking solution quickly to each slide then place a plastic coverslip over the solution and incubate at room temperature for 5 min.

3   Peel off the plastic coverslip gently and apply 60 μl of FITC–avidin to each slide; cover the solution by a piece of new plastic coverslip.[b]

4   Incubate the slides at 37 °C for 20 min in a humidified chamber with a light-protected cover.

5   Remove the coverslips and rinse the slides in washing solution C at room temperature three times for 3 min each.

6   Apply 60 μl of blocking solution to each slide. Cover the solution with a plastic coverslip and leave the slide at room temperature for 5 min.

Protocol 11 continued

7 Add 60 μl of biotinylated goat anti-avidin antibody to each slide. Replace the coverslip and incubate at 37 °C for 20 min.

8 Remove the coverslip and transfer the slides into washing solution C, three times for 3 min each.

9 Apply 60 μl of blocking solution to each slide. Place a plastic coverslip over the solution and incubate at room temperature for 5 min.

10 Peel off the plastic coverslip and apply 60 μl of FITC avidin to each slide. Cover the solution by a piece of new plastic coverslip.

11 Incubate the slides at 37 °C for 20 min in a humidified chamber with a light-protected cover.

12 Remove the coverslips and rinse the slides in washing solution C, and then 2 × SSC at room temperature for 3 min each.

[a] More than one fluorochrome is often used for fibre FISH to distinguish two probes. The two most popular are FITC and TRITC. For example, if biotin labelling is used for one probe and detected by avidin–FITC, it can be amplified by using biotinylated goat anti-avidin, and then detected by one more layer of avidin–FITC. If digoxigenin labelling is used for another probe, anti-digoxigenin–rhodamine Fab fragments can be used for detection. Alternatively, the amplification and detection can be achieved by using mouse anti-digoxigenin, then rabbit anti-mouse-rhodamine (TRITC), or goat anti-rabbit–TRITC.

[b] It is very important to keep the slides away from light from this step onward.

6.5 Counterstaining and antifade

Protocol 12

Counterstaining and antifade

Reagents

- DAPI (Sigma): 0.2 mg/ml of stock solution in H$_2$O; store in the dark at 4 °C

- Propidium iodide (PI, Sigma): 0.1 mg/ml of stock solution in PBS (phosphate-buffered saline) (Invitrogen)

- 2 × SSC (before and after 4′,6-diamidino-2-phenylindole [DAPI] staining wash)

- Antifade solution: dissolve 100 mg of p-phenylendiamine in 10 ml of PBS, then add 90 ml of glycerol. Adjust the pH to 9.0 by using litmus paper. Store at −20 °C.

Method

1 Immerse the slides in 0.2 μg/ml DAPI in 2 × SSC at room temperature for 5 min. Rinse them in 2 × SSC three times for 1 min each.

2 Overlay the sample with 5 μl PI mixed with 10 μl of antifade solution and cover it with a 22 × 40 mm coverslip.

3 Gently press the coverslip to exclude excessive liquid.

Note: For two-colour detection involving rhodamine the propidium iodide (PI) will be avoided.

7 Photography

The slides may be examined immediately or stored at –20°C for one to two days. Use the DAPI filter to identify DAPI stained chromatin or DNA fibres first, then switch to the FITC or rhodamine or dual filter to localize the FISH signals. An example of a fibre FISH is shown in *Figure 1* (see also *Plate 7*). Specialized equipment such as a sensitive charged-couple device (CCD) camera is needed for fibre FISH. In many cases, quantitative analysis and image processing is required. If these systems are not available, high-speed film should be used to record the images. Kodak Ektachrome p800/1600 can be push-developed up to 3200.

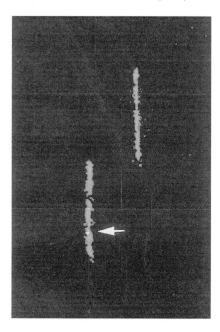

Figure 1 (See *Plate 7*) Fibre FISH analysis examining a heterozygous transgene in a specific genomic region. The green signal (arrow) shows the locus of the transgene within the region highlighted by the BAC probe (red signal).

Acknowledgements

Free chromatin fibre FISH was originally developed in Dr Lap Chee Tsui's Laboratory at the Hospital for Sick Children, Toronto. I thank him for his encouragement and support. A special thanks to Drs Steven Bremer and Christine Ye for editing this manuscript. This work was supported by the Center for Molecular Medicine and Genetics (Wayne State University School of Medicine) start up fund.

References

1. Heng, H. H. Q. and Tsui, L.-C. (1998). *J. Chromatogr. A*, **806**, 219.
2. Lichter, P., Tang, C. C., Gall, K., Hermanson, G., Evans, G., Housman, D., *et al.* (1990). *Science*, **247**, 64.
3. Trask, B., Pinkel, D., and Engh, G. V. D. (1989). *Genomics*, **5**, 710.
4. Moens, P. and Pearlman, R. E. (1990). *Cytogenet. Cell Genet.*, **53**, 219.

5. Brandriff, B., Gordon, L., and Trask, B. (1991). *Genomics*, **10**, 75.

6. Heng, H. H. Q., Squire, J., and Tsui, L.-C. (1991). *Proc. 8th Int. Cong. Hum. Gen. Am. J. Hum. Genet. Suppl.*, **49**, 368.

7. Heng, H. H. Q., Squire, J., and Tsui, L.-C. (1992). *Proc. Natl. Acad. Sci. USA*, **89**, 9509.

8. Wiegant, J., Kalle, W., Mullenders, L., Brookes, S., Hovers, J. M. N., Dauwerse, J. G., *et al.* (1992). *Hum. Mol. Genet.*, **1**, 587.

9. Parra, I. and Windle, B. (1993). *Nature Genet.*, **5**, 17.

10. Fidlerova, H., Senger, G., Kost, M., Sanseau, P. M., and Sheer, D. (1994). *Cytogenet. Cell Genet.*, **65**, 203.

11. Houseal, T. W., Dackowski, W. R., Landes, G. M., and Klinger, K. W. (1994). *Cytometry*, **15**, 193.

12. Haaf, T. and Ward, D. C. (1994). *Hum. Mol. Genet.*, **3**, 697.

13. Weier, H.-U. G., Wang, M., Mullikin, J. C., Zhu, Y., Cheng, J.-F., Greulich, K. M., *et al.* (1995). *Hum. Mol. Genet.*, **4**, 1903.

14. Heiskanen, M., Karhu, R., Hellsten, E., Peltonen, L., Kallioniemi, O. P., and Palotie, A. (1994). *BioTechniques*, **17**, 928.

15. Michalet, X., Ekong, R., Fougerousse, F., Rousseaux, S., Schurra, C., Hornigold, N., *et al.* (1997). *Science*, **277**, 1518.

16. Florijn, R. J., Bonden, A. J., Vrolijk, H., Wiengant, J., Vaandrager, J.-W., Baas, F., *et al.* (1995). *Hum. Mol. Genet.*, **4**, 831.

17. Heng, H. H. Q. (2000). In *Methods in molecular biology* (ed. I. A. Darby), Vol. 123, p. 69. Humana Press, Totowa, NJ.

18. Heng, H. H. Q., Spyropuulos, B., and Moens, P. B. (1997). *Bioessays*, **19**, 75.

19. Fransz, P. F., Alonso-Blanco, C. M., Liharska, T. B., Peeters, A. J. M., Zabel, P., and de Jong, J. H. (1996). *Plant J.*, **9**, 421.

20. Jackson, S. A., Wang, M. L., Goodman, H. M., and Jiang, J. (1998). *Genome*, **41**, 566.

21. Kuitunen, N. H., Aalotonen, J., Yaspo, M.-L., Eeva, M., Wessman, M., Peltonen, L., *et al.* (1999). *Genome Res.*, **9**, 62.

22. Heiskanen, M., Peltonen, L., and Palotie, A. (1996). *Trends Genet.*, **12**, 379.

23. Dunham, I., Shimizu, N., Roe, B. A., Chissoe, S., Hunt, A. R., Collins, J. E., *et al.* (1999). *Nature*, **402**, 489.

24. Shiels, C., Coutelle, C., and Huxley, C. (1997). *Genomics*, **44**, 35.

25. Heng, H. H. Q., Chamberlain, J. W., Shi, X.-M., Spyropoulos, B., Tsui, L.-C., and Moens, P. B. (1996). *Proc. Natl. Acad. Sci. USA*, **93**, 2795.

26. Lestou, V. S., Strehl, S., Lion, T., Gadner, H., and Ambros, P. F. (1996). *Cytogenet. Cell Genet.*, **74**, 211.

27. Raderschall, E., Golub, E. I., and Haaf, T. (1999). *Proc. Natl. Acad. Sci. USA*, **96**, 1921.

28. Pandita, A., Godbout, R., Zielenska, M., Thorner, P., Bayani, J., and Squire, J. A. (1997). *Genes Chromosomes Cancer*, **20**, 243.

29. van Ommen, G. J., Breuning, M. H., and Raap, A. K. (1995). *Curr. Opin. Genet. Dev.*, **5**, 304.

30. Heng, H. H. Q. and Tsui, L.-C. (1994). In *Methods in molecular biology* (ed. K. H. A. Choo), Vol. 33, p. 109. Humana Press, Totowa. NJ.

31. Heng, H. H. Q., Tsui, L.-C., Windle, B., and Parra I. (1995). In *Current protocols in human genetics* (ed. N. Dracopoli, J. Haines, B. Korf, D. Moir, C. Morton, C. Seidman, J. Seidman, and D. Smith), p. 4.5.1. John Wiley and Sons, New York.

32. Heng, H. H. Q., Chen, W. Y., and Wang, Y. C. (1988). *Mutat. Res.*, **199**, 199.

33. Heng, H. H. Q. and Shi, X. M. (1997). *Cell Res.*, **7**, 119.

34. Heng, H. H. Q. and Tsui, L.-C. (1993). *Chromosoma*, **102**, 325.

35. Heng, H. H.Q. and Tsui, L.-C. (1994). In *Methods in molecular biology* (ed. K. H. A. Choo), Vol. 33, p. 35. Humana Press, Totowa, NJ.

Chapter 6

Applications of RNA FISH for visualizing gene expression and nuclear architecture

Rose Tam, Lindsay S. Shopland, Carol V. Johnson, John A. McNeil, and Jeanne B. Lawrence
Department of Cell Biology, University of Massachusetts Medical School, 55 Lake Avenue North, Worcester, MA 01655, USA.

1 Introduction

Fluorescence *in situ* hybridization is not a singular technique, but a versatile experimental approach that provides ever more precise molecular information directly in the context of cellular structure. This chapter will focus on one major aspect of the molecular cytological approach: the detection of nuclear RNAs, together with their genes or other biochemical components of the cell. Improvements in the development and application of FISH in recent years have demonstrated the value of this approach for studying gene expression as well as processing, transport, and function of various cellular and viral RNAs. Although the basic technique described here is also applicable to detection of cytoplasmic mRNA, our focus will be on methods optimized primarily for high resolution, high sensitivity detection of nuclear transcripts, in precise relation to other nuclear components.

Standard molecular biology methods typically involve analysis of nucleic acids extracted from cells; therefore, they are limited by design to an averaged result obtained over a population of cells, often not physiologically homogeneous. Perhaps the most common use of RNA *in situ* hybridization is to assess gene expression on a single cell basis. If the goal is simply to assess which cells in a population are expressing a given RNA, then any *in situ* hybridization method which can detect that RNA within individual cells can be used, including autoradiographic or colorimetric methods. However, the detection of nuclear RNA by FISH can in some cases be advantageous for assessing expression of a gene within individual cells. As illustrated in *Figures 1* and *2* (see also *Plates 8* and *9*), the expression of myosin heavy chain RNA and collagen type I α1 (COL1A1) RNA are readily detected in the cytoplasm and nucleus of muscle fibres and human fibroblasts, respectively. However, the dystrophin RNA (*Figure 1*; see also *Plate 8*,

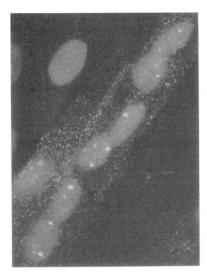

Figure 1 (See *Plate 8*) Detection of high and low abundance transcripts in nuclei versus cytoplasm of skeletal muscle fibres. Male human multinucleated myotubes were hybridized to detect muscle-specific myosin heavy chain (MyHC) and dystrophin RNAs. Most nuclei have two MyHC RNA foci (green) and a single dystrophin (on the X chromosome) RNA signal (red). While MyHC mRNA is evident in the myotube cytoplasm (green), the extremely low levels of dystrophin RNA in the cytoplasm are difficult to detect by FISH. Dystrophin gene expression is easily detected due to the large focal accumulation of nuclear RNA at the gene. A single undifferentiated myoblast nucleus (upper left) express neither MyHC nor dystrophin RNA (1).

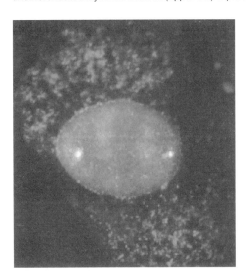

Figure 2 (See *Plate 9*) Visualization of mRNA distribution from gene to cytoplasm. Although procedures here are optimized for nuclear detection, in many cases nuclear and cytoplasmic transcripts can be detected simultaneously. In the human fibroblast shown here, sequential hybridizations to collagen type I α1 (COL1A1) RNA (red) and then the COL1A1 gene (green) reveals nuclear RNA foci emanating from the gene (overlap appears yellow), and distributed in the cytoplasm. More dispersed nuclear signal likely represents transcripts moving towards the nuclear envelope (19).

red), which is present at very low levels in the cytoplasm, is most easily detected at its site of synthesis in the nucleus, even using a probe that detects less than 1% of the full-length primary transcript (1). Although not widely appreciated, the most sensitive detection for some mRNAs will be derived from fluorescence hybridization to the nuclear RNA, since this can be the site of highest concentration in the cell, providing a discrete signal clearly distinguishable from background.

Although methods presented here will delineate which cells in a population express a given transcript, they can also be used for more powerful applications that put far greater demands on the experimental approach. For example, we may want to know how many or which alleles *within the same* nucleus are being expressed (*Figure 3*; see also *Plate 10*), or compare the processing or transport of transcripts from a normal and mutant allele within an individual heterogeneous cell (*Figure 7C* and *D*; see also *Plate 13C* and *D*). Such experiments require an approach that can deliver the best combination of extremely high sensitivity, resolution, and specificity, while being flexible enough to simultaneously label multiple DNA/RNA–protein components. Hence, the techniques summarized here will be of greatest value for studies in which the *intracellular* localization of RNA, not just its presence within the cell, is important for the biological or clinical questions investigated.

In this chapter, recent innovations and improvements to our standard laboratory procedures, and examples as applied to individual genes and transcripts are presented. The specific examples illustrated here derive from our recent work using RNA–DNA FISH in combination with immunocytochemistry to study both X chromosome inactivation and the relationship of gene expression and RNA metabolism to nuclear structure. In addition to basic science questions such as these, the ability to detect and localize small quantities of specific nucleic acids in single cells clearly has relevance for a variety of clinical applications. For a broader treatment of the development and applications of DNA/RNA *in situ*

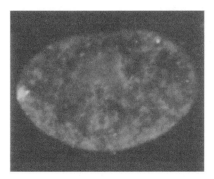

Figure 3 (See *Plate 10*) Nuclear RNA FISH can determine which alleles within a single cell are active. The UBE-1 gene is one that escapes inactivation on the inactive X chromosome. In contrast, the human XIST gene is expressed solely from the inactive X chromosome, and produces a large RNA accumulation (red) that 'paints' the inactive X. This contrasts with the much smaller UBE-1 nuclear RNA foci (green), more typical of signal for protein coding genes (7).

95

hybridization methodology, the reader is referred to previous reviews on this subject (2–4).

2 Cell preparation

The choice of a cell preparation protocol is determined primarily by the type of cell sample, the target RNA, and the objective of the experiment. To detect sequences within the nucleus, the protocol must permeabilize the nucleus sufficiently to allow the probe access to its target and, at the same time, fix the cell sufficiently to preserve labile RNAs and the integrity of structure. Many monolayer cells have extensive cytoplasm that requires extraction in detergent prior to fixation in order to adequately permeabilize the cell nucleus and allow detection of nuclear targets, hence the protocol detailed below uses this procedure. The detergent extraction method does not appear to significantly diminish or disrupt nuclear RNA, based on comparisons of RNAs that can be detected with or without the extraction (5–7). Because RNA is most vulnerable to degradation when cells are permeabilized prior to fixation, it is desirable to minimize the duration of pre-fixation steps. Often cytoplasmic transcripts can also be detected after 1–5 min of Triton treatment before fixation (*Figure 2*; see also *Plate 9*). However, these longer Triton extractions risk the loss of cytoplasmic mRNA and, if detection of cytoplasmic mRNA is the main goal, procedures using either very short (20–30 sec) Triton treatment before fixation, or longer Triton treatment *after* fixation, are preferable. We have also had success with the cell preparation protocol developed in Thomas Cremer's lab for DNA hybridization, which involves Triton and saponin extraction after fixation (8) (see also Chapter 7). Upon RNA hybridization, this protocol yields results that are very similar to cells extracted before fixation. The consistency of the results between the two quite different protocols supports the conclusion that extracting before fixation does not perturb the distribution of RNA in the nucleus.

It is important to note that not all cell types require permeabilization with detergent. We have found several cell types grown in suspension, which tend to have thinner cytoplasm, and require no extraction to detect nuclear sequences (e.g. lymphoma cells or human lymphocytes) (9, 10). These cells can simply be gently placed on to a multiwell slide (Cel-line) at the proper cell density and allowed to air dry briefly prior to fixation with formaldehyde (see below). Suspension cells can also be cytospun on to glass slides (11). This procedure flattens rounded cells, which is advantageous for detecting weaker signals, but less ideal for preserving cell morphology. Detergent extraction is also generally unnecessary in tissue sections, since nuclei at the surface of the tissue are often exposed during cutting of the section.

2.1 Detergent-extracted cell preparation

The standard protocol below is optimized for detecting nuclear RNA in monolayer cells (such as fibroblasts, epithelial cells, or skeletal muscle) and also works well for detecting DNA and most nuclear proteins.

Protocol 1

Extraction of monolayer cells

Equipment and reagents

- No. 1.5 coverslips
- Coplin jar
- Hanks' balanced salt solution (Invitrogen)
- 0.5% Triton X-100 in CSK buffer
- 200 mM vanadyl ribonucleoside complex (VRC, Invitrogen)
- Cytoskeleton (CSK) buffer:[a] 100 mM NaCl, 300 mM sucrose, 10 mM piperazine-N,N'-bis(2-ethane sulfonic acid) (Pipes) pH 6.8, 3 mM $MgCl_2$, 1.2 mM phenylmethylsulfonyl fluoride
- 4% paraformaldehyde in PBS (Ted Pella)
- 70% ethanol

Method

1 One to three days prior to fixation, grow cells to subconfluency on the coverslips.

2 Transfer coverslips to a Coplin jar and rinse several times with Hanks' balanced salt solution to remove media.

3 Rinse coverslips with CSK buffer at room temperature.

4 Gently extract cells with 0.5% Triton X-100 in CSK buffer for 1–5 min on ice. More structured cells, such as skeletal muscle fibres, require longer (e.g. 5 min) extraction. To prevent degradation of cellular RNA, include VRC in the extraction buffer at a final concentration of 10 mM. This may be crucial for detecting low copy transcripts.

5 Immediately fix cells for 10 min in 4% formaldehyde in PBS.

6 Rinse in 70% ethanol.

7 Cells can be stored in 70% ethanol at 4 °C for several weeks. Prolonged storage may cause RNA degradation. Ethanol can interfere with immunodetection of some proteins, which must be determined empirically. In such cases, cells can be stored for a few days in PBS.

[a] From methods in ref. 12.

2.2 Cytogenetic preparations

In some cases, it is desirable to examine RNA in a cytogenetic preparation where mitotic chromosomes can be evaluated with interphase nuclei. Cytogenetic preparations use methanol/ acetic acid fixation, which often does not preserve RNA well (13). In this case, RNA retention depends on properties of the individual transcript. For example, Epstein–Barr virus transcripts and XIST RNA are stable enough to be retained following a cytogenetic cell preparation, but transcripts from most cellular genes are not (7, 9).

A standard *in situ* harvest technique for cytogenetic cell preparations (14) has been adapted by K. Wydner, M. Byron, and C. Clemson. Using cells that have a high mitotic index bypasses the need to arrest cells in metaphase with colcemid. BrdU incorporation is included to elongate chromosomes and enhance banding.

Other cell preparation techniques that are aimed at more specialized structures including nuclear matrix (6) and nuclear halo preparations (15) are described in detail elsewhere (see also ref. 16).

Protocol 2

Preparation of cytogenetic cells and chromosomes

Equipment and reagents

- Tissue culture centrifuge
- Glass slides
- Steaming water-bath
- Slide warmer
- 100 mM BrdU (1000 × stock)
- Trypsin/EDTA: 0.05% trypsin, 0.53 mM EDTA (Invitrogen)
- Hanks' balanced salt solution
- Hypotonic solution: 75 mM KCl
- Fixative: 1:3 ratio of acetic acid/methanol

Method

1 4 h prior to harvest, add BrdU to the culture medium to a final concentration of 0.1 mM.

2 In the case of monolayer cells, trypsinize and wash cells by centrifuging at 150 g for 5 min and resuspending the pellet with 5 ml Hanks' buffer. Repeat the wash.

3 Resuspend cells in 1 ml of Hanks' buffer and gently add 7 ml of hypotonic solution. Incubate at 37 °C for 10 min.

4 Pellet by centrifugation at 150 g for 5 min.

5 Remove most of the hypotonic solution, leaving 1 ml with cell pellet. Completely resuspend the cells in this.

6 Add 1 ml of the fixative drop by drop. Continue to add more fixative to a final volume of 6 ml.

7 Pellet by centrifugation at 150 g for 5 min.

8 Repeat steps 6 and 7 four more times.

9 Resuspend cells to desired density in fixative and drop onto cleaned glass slides from a high point above slides, preferably in a humid environment for optimal spreading of chromosomes.

10 Pass each slide over a steaming water-bath and air dry on a slide warmer.

3 Probe preparation

The choice of DNA sequence used to make a probe can greatly affect the outcome of hybridization *in situ*. The size of the target sequence and the choice of whether a cDNA or a genomic DNA sequence is used will influence the intensity of the hybridization signal. Use of a cDNA probe is advantageous for detection of cytoplasmic mRNAs, however, the choice is more complicated for nuclear RNAs. The larger genomic sequence, with its many introns and flanking sequences, is more efficient for DNA hybridization or for detection of unspliced primary transcripts. We can detect gene signals with probes targeting as little as 1–2 kb in interphase nuclei and metaphase chromosomes, and specific nuclear RNA signals have been detected with oligonucleotides as small as 22 bp (20). Ideally, probes targeting 5–10 kb or more work best for single copy gene detection. The presence of repetitive elements in the probe is not a problem, provided Cot-1 DNA is added to compete these sequences (below). Likewise, with cloned plasmid DNA, it is not necessary to remove the inserted sequence of interest from the rest of the vector sequences.

Most commonly our probes are labelled by incorporating biotin or digoxigenin nucleotide analogues on to double-stranded DNA through a nick translation reaction. Detection of digoxigenin probes is more sensitive than that of biotin probes, but also generally produces more background. Endogenous biotin in some cell types can cause a significant problem with use of biotin labelled probes, but this difficulty is usually circumvented by the detergent extraction of the cell (above).

At the relatively high probe concentrations necessary for single gene detection (17), non-specific sticking of the probe can become more pronounced (discussed in ref. 16). Longer probe fragments tend to self-associate and form aggregates during the hybridization reaction, which can cause prohibitive background. For hybridization *in situ*, generating small probe fragments of less than 200 nucleotides in length is key for avoiding high background and optimizes penetration of cell structure.

Fragment size can be adjusted by varying the concentration of enzymes (DNase I and DNA polymerase) in the nick translation reaction. We have found it convenient to use a commercially available mixture of nick translation enzymes that are already calibrated and quality controlled for this purpose. The protocol detailed below uses the enzyme mix from Invitrogen BioNick Labelling Kit. The buffer/nucleotide mix from this kit is not used because it contains a mixture of biotin–dATP and 'cold' dATP, whereas we have found that inclusion of 100% labelled nucleotide results in the strongest probe signals. In addition, the efficiency of the reaction depends on the purity of the template DNA; CsCl preparations work consistently well in our hands.

Protocol 3

Nick translation of DNA probes

Equipment and reagents

- 20 °C water-bath
- Cesium chloride grade DNA
- 10 × nick translation buffer (NTB): 0.5 M Tris pH 7.5, 0.1 M $MgSO_4$, 0.5 mg/ml RNase-free BSA (Roche Diagnostics), 1 mM dithiothreitol
- 1 mM labelled dUTP (either biotin-16-dUTP or digoxigenin-11-dUTP) (Roche Diagnostics)
- Nucleotide mix of 600 μM each of dATP, dCTP, and dGTP (Roche Diagnostics)
- Enzyme mix from Invitrogen BioNick DNA Labelling System
- 5% sodium dodecyl sulfate (SDS)
- 0.5 M EDTA
- Salmon sperm DNA (Sigma)
- Ethanol

Method

1 For nick translating 0.5 μg of DNA, which serves for ten hybridizations, mix the following reagents in a microcentrifuge tube:
   - 2.5 μl of 10 × NTB
   - 2.5 μl nucleotide mix
   - 3.0 μl of 1 mM labelled dUTP
   - 0.5 μg DNA to be labelled
   - H_2O for a final reaction volume of 25 μl
   - 2.5 μl enzyme mixture

2 Incubate the reaction for 2.5 h at approx. 20 °C.

3 Add 2.5 μl of 5% SDS and 2.5 μl of 0.5 M EDTA.

4 Heat inactivate the reaction by incubating for 10 min at 65 °C.

5 Add 5 μg salmon sperm DNA.

6 Ethanol precipitate.

7 Wash pellet with 70% ethanol. (Omitting this step will result in probe with high background.)

8 Resuspend in 100 μl H_2O for a final concentration of ~5 ng/μl probe.

9 Probe can be stored at –20 °C for at least a year.

4 Hybridization to RNA

In addition to the quality of the cell preservation and probe preparation, there are several elements that are also important for successful hybridization to RNA. The quality of reagents used for hybridization (for example the brand of formamide) should be of high molecular biology grade and free of any RNase. Our laboratory routinely hybridizes cells overnight for convenience. However, because the probe concentration recommended is generally in excess, hybridization times as short as 3 h are successful and may be advantageous to minimize cell/RNA degradation. Furthermore, the addition of VRC (see *Protocol 1*) to the hybridization mixture inhibits RNase activity.

In order to reduce cross-hybridization to highly repetitive sequences, excess unlabelled Cot-1 DNA is generally added to the probe mixture prior to hybridization, as are other non-specific nucleic acid competitors. These have proven to be critical for limiting sequence detection to specific RNA sequences. Cot-1 DNA repetitive sequences are species specific, thus the Cot-1 DNA used for hybridization should be derived from the same species as the target cells.

Hybridized probes can be indirectly detected with commercially available avidin or anti-digoxigenin antibodies that have been conjugated to Texas Red, rhodamine, or FITC (Roche). Alexa dyes conjugated to secondary antibodies or avidin (Molecular Probes) are another source for secondary detection and may have the advantages of being more sensitive, stable, and resistant to photobleaching. Each lot of secondary detector can vary in terms of titre and label content. Therefore, a range of concentrations should be tested initially to determine optimal conditions.

4.1 Basic RNA hybridization procedure

Protocol 4

Basic RNA hybridization procedure

Equipment and reagents

- Speed Vac lyophilizer
- Heat block at 80 °C
- 37 °C incubator
- Sheets of Parafilm
- Ethanol series: 70%, 95%, and 100%
- Nick translated probe (see *Protocol 3*)
- Competitor nucleic acid sequences: 10 μg each of sonicated salmon sperm DNA, *E. coli* tRNA, and Cot-1 DNA
- Formamide (ACS reagent grade, Sigma)
- Hybridization buffer: 0.4% BSA, 4 × SSC, 20% dextran sulfate, 40 mM VRC

- 20 × SSC stock: 3 M NaCl, 0.3 M sodium citrate pH 7.4
- Fluorochrome-conjugated secondary detection (either avidin or anti-digoxigenin antibody)
- RNase-free BSA (Roche Diagnostics)
- 10% Triton X-100 (Roche Diagnostics)
- 200 mM VRC (Invitrogen)
- Mounting media:[a] 0.1% phenylenediamine, 90% glycerol, PBS adjusted to pH 9.0 with sodium bicarbonate

Method

1 Dehydrate the cells on a coverslip through an ethanol series, air dry, and reserve for step 5, or remove from ethanol storage and rehydrate in 1 × PBS.

2 Lyophilize 50 ng of nick translated probe with 10 μg of each competitor sequence.

101

3 Resuspend the probe in 10 μl of 100% formamide (Sigma) and heat denature at 80 °C for 10 min.

4 Add 10 μl of hybridization buffer to the heat denatured probe.

5 Immediately spot probe mix onto a sheet of Parafilm, overlay with the coverslip (cell side down), and seal with a second sheet of Parafilm to prevent evaporation. Incubate at 37 °C in a humidified chamber for 3 h to overnight.

6 Rinse the coverslip in the following sequence for at least 15 min at each step:
   (a) 50% formamide, 2 × SSC at 37 °C.
   (b) 2 × SSC at 37 °C.
   (c) 1 × SSC at 25 °C.

7 Equilibrate in 4 × SSC for 1–2 min.

8 Detect the signal by incubating in either fluorochrome-conjugated biotin or fluoro-chrome-conjugated anti-digoxigenin antibody at a concentration of 3 μg/ml in 4 × SSC, 1% BSA, 40 mM VRC for 20 min–1 h at 37 °C.

9 Rinse for 10 min through each step:
   (a) 4 × SSC.
   (b) 4 × SSC, 0.1% Triton X-100.
   (c) 4 × SSC.

10 At this point, the signal may either be fixed for 10 min in 4% paraformaldehyde, PBS in preparation for subsequent procedures (see *Protocol 7*), or else it may be mounted and sealed for viewing. Cellular DNA can be counterstained with 4′,6-diamidino-2-phenylindole (DAPI) prior to mounting. To mount, add a drop of mounting media and seal the slide with nail polish.

[a] Commercial sources of pre-made mounting media are also available from Vysis and Vector Laboratories.

4.2 Oligonucleotide hybridization

Although targeting larger sequences where possible is preferable because their higher complexity reduces unwanted noise, especially when detecting DNA, oligonucleotide probes can be used for some RNA detection strategies. Oligos can be made or purchased (Operon, Biosource International) with a variety of labels (e.g. fluorescein, biotin) in varying numbers and positions. For some experiments, we have had success using 20–25 nt long oligos with a single biotin label at the 3′ end. The oligo should be HPLC purified for best results. These oligos have worked well for the detection of U2 and pre-U2 RNA in the nucleus (*Figure 4*) (20), as well as for detecting specific collagen introns (*Figure 11*; see also *Plate 16*) (21). However, differences in sequence content can greatly affect the hybridization and background properties of an oligonucleotide. While these probes can provide excellent specificity, their potential for spurious results or false hybridization should be well-controlled in the experimental design.

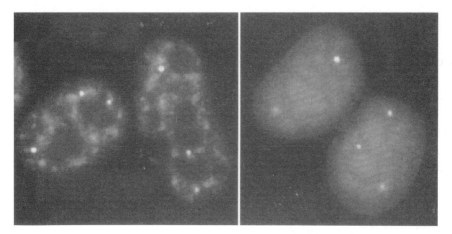

Figure 4 Oligonucleotide probes provide the high specificity needed to discriminate pre-U2 RNA from mature U2 RNA. The U2 gene produces a 199 nt long pre-U2 RNA, 11 nucleotides of which are removed to create the mature U2 snRNA. A 22 nt long 3′ biotinylated oligonucleotide probe which is complimentary to nt 29–50 of U2 snRNA detects U2 in Cajal (coiled) bodies (bright foci) and splicing factor-rich speckles in HeLa cell nuclei (left panel). In contrast, an oligonucleotide of the same length complimentary to 15 nt of the mature U2 and 7 nt of the pre-U2 tail that are removed from the mature U2 detects only RNA in the Cajal body, under identical hybridization conditions (right panel). A negative control oligonucleotide with 15 nt of U2 sequence and 7 nt random sequence showed no specific signal (20).

Oligo hybridizations are performed using *Protocol 4* with minor modifications: a lower quantity of probe (5–20 ng) and a lower concentration of formamide (10–20%) in the hybridization buffer and in the wash. The formamide concentration must be adjusted for optimal efficiency depending on the length and GC content of the oligonucleotide probe. Competitor DNA is still included in the hybridization probe mix (10 μg each of ssDNA, tRNA, and Cot-1), and therefore the probe mix must be heat denatured before application to the slide. Another oligo hybridization protocol has been reported (18) and works equally well in our hands.

5 Hybridization to DNA

We have demonstrated several methods to target specifically and discriminate between RNA and DNA hybridization *in situ* (9, 16). The RNA hybridization procedure above precludes cross-hybridization to the gene, because essentially no hybridization to cellular DNA is detected if the double-stranded cellular DNA has not been denatured (9). In contrast, when the goal of the experiment is to strictly pinpoint the gene signal, residual RNA should be removed to prevent cross-hybridization.

If the gene of interest is transcriptionally active in cells studied, RNA can be removed by NaOH hydrolysis, RNase A pre-treatment, or RNase H post-hybridization (9, 19). NaOH treatment is the simplest and possibly most thorough method and has the advantage that it simultaneously denatures DNA for subsequent hybridization. Hybridization to mitotic cells also provides a good control to verify DNA versus RNA signals, since these cells are not transcriptionally active, hence localized FISH signals represent DNA. To ensure hybridization is to RNA, either mitotic cells or a cell type that does not express the gene can be used as a negative control.

Treatment with RNase or NaOH does not necessarily remove RNA effectively or totally, although it is often assumed that it does. To control for the effective elimination of RNA when only DNA signal is desired, a good control is to omit denaturation of cellular DNA. If RNA has been effectively eliminated, this control should be completely negative.

5.1 Detecting heat denatured cellular DNA

Unlike detection of cellular RNA where preservation of the nucleic acid is key, detection of cellular DNA requires effective denaturation of double-stranded DNA. Our standard protocol uses heat denaturation, although when called for experimentally, we have used NaOH denaturation with variable success.

Protocol 5

Heat denaturation of DNA

Equipment and reagents

- Coplin jar
- Microwave oven
- 70 °C water-bath
- Thermometer

- 20 × SSC stock: see *Protocol 4*
- Formamide (ACS reagent, Sigma)
- Ice-cold ethanol series: 70%, 95%, 100%

Method

1 Equilibrate cells fixed on coverslips in 2 × SSC.

2 Microwave heat a solution of 70% formamide, 2 × SSC in a Coplin jar to 80 °C and transfer to a water-bath set at 70 °C. It is important not to allow the solution to sit at high temperatures for extended periods of time, as the pH of the solution may be altered and affect hybridization.

3 When the temperature is at exactly 74 °C, transfer the coverslip to the denaturation solution. Incubate for exactly 2 min. (The temperature should drop slightly but stay as close to 70 °C as possible.)

4 Transfer immediately to a Coplin jar containing ice-cold 70% ethanol and dehydrate through a 95% and 100% cold ethanol series.

5 Air dry.

6 Hybridize to denatured DNA according to *Protocol 4*. (VRC can be omitted from the hybridization buffer when preservation of RNA is not the objective.)

5.2 DNA FISH using NaOH denaturation and RNA hydrolysis

NaOH denaturation has worked well in our laboratory but has not proven as consistent as heat denaturation. Success at removing cellular RNA by this method also varies, depending both on the specific RNA target and the concentration of NaOH.

Protocol 6

Alkaline denaturation of DNA

Equipment and reagents

- Coplin jar
- Alkaline solution: 0.07 M to 0.2 M NaOH in 70% ethanol
- 70% ethanol
- 100% ethanol

Method

1 Incubate the coverslip in alkaline solution for 5 min.

2 Wash the coverslip for 5 min in 70% ethanol.

3 Repeat the wash in 100% ethanol.

4 Air dry.

5 Hybridize according to *Protocol 4*. (VRC may be omitted.)

---

6 Multiple label techniques and applications

6.1 Coupling the detection of RNA with DNA

Experiments using multiple labels have innumerable applications for studying the spatial relationships between different cellular components. Distinguishing specific genes from their transcripts within the same nucleus has proven especially valuable for investigating the spatial organization of RNA metabolism. We have adapted our FISH protocols to visualize RNA in one colour and DNA in another. Using this technique, we have observed accumulations of transcripts from a few specific genes, for example COL1A1 and myosin heavy chain (MyHC), that extend in one direction away from the gene, suggesting vectorial transport of those RNAs away from their sites of transcription (*Figure 5*; see also *Plate 11*). This procedure will work with probes derived from the same sequence as long as they are differentially labelled. These probes are hybridized sequentially to differentiate between RNA and DNA as described in *Protocol 7*.

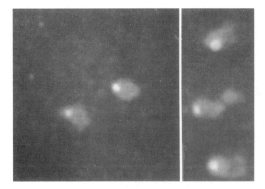

Figure 5 (See *Plate 11*) Simultaneous visualization of a gene and RNA emanating from that gene, shown for MyHC (right) and COL1A1 (left). Visualization of nuclear RNA (red) in direct spatial relation to the gene (green), can determine not only if both alleles are active, but also identify structural details with potentially important biological implications. Results of sequential RNA and DNA hybridizations (19) with a differentially labelled cDNA sequence were visualized using a dual bandpass filter (to avoid any optical shift). The two signals are not simply coincident, but rather the COL1A1/MyHC RNA 'tracks' (red) are larger than the DNA signal and consistently positioned to one side of it (1, 19, 21). Often the prominent RNA signals are somewhat elongated and extend 0.5–3 μm beyond the gene (20–30 kb), which itself is at or below the resolution of fluorescence microscopy (0.2 μm). What may seem small differences in size and position by light microscopy are actually quite large and significant on a molecular scale.

Protocol 7

Sequential detection of RNA and DNA

Equipment and reagents

- See *Protocols* 4 and 6

Method

1   Hybridize first to RNA according to *Protocol 4* using one of the forms of the labelled probe (e.g. digoxigenin) and detecting it in one colour (e.g. rhodamine).

2   Fix RNA signal by incubating in 4% paraformaldehyde, PBS for 10 min.

3   Wash twice, 5 min each, with PBS.

4   Follow *Protocol 6* for alkaline denaturation of DNA and hybridize with a probe containing another type of label (e.g. biotin) and detecting the probe in a second colour (e.g. FITC).

5   If cellular protein will also be detected, fix DNA signal as in step 2, wash as in step 3, and proceed with *Protocol 8*. Otherwise, mount cells for viewing. ·

6.2 Coupling protein detection with FISH

Double- and triple-label experiments that combine RNA and/or DNA FISH with protein immunolocalization are invaluable for examining structural relationships between different genes or RNAs and other biochemical components of the cell. We have linked together hybridization and immunofluorescence techniques to determine whether specific genes or transcripts spatially associate with certain nuclear compartments or structures. In cases where they do, we have extended the technique to determine the precise alignment of gene, transcript, and nuclear body in triple-label experiments. Such triple-label experiments have shown, for example, that the accumulation of transcripts from the U2 snRNA locus does not necessarily mediate the preferential interaction between that gene and the Cajal body (*Figure 10*; see also *Plate 15*) (20).

Like the coupling of DNA and RNA FISH, linking the immunodetection of proteins with FISH occurs in sequential steps and can be affected by the order of the steps. We have found that some epitopes are destroyed by prior hybridization. In these cases, cells can be immunostained first and subsequently fixed to preserve that signal. If RNA detection is to follow, the addition of VRC to buffers during incubations with antisera helps prevent RNA degradation. Note also that some epitopes are compromised by exposure to ethanol, in which case cells can be stored in PBS instead but should be used within one to two days after fixation.

Triple-label experiments require the use of three distinct fluorophores. In addition to the more common fluorescein (FITC) and rhodamine or Texas Red agents (TRITC), we often use the coumarin derivative, aminomethylcoumarin (AMCA). AMCA-conjugated secondary antibodies can be used for labelling abundant protein antigens, but due to its low coefficient of extinction, AMCA is generally not intense enough to label transcripts or gene signals. In multiple label experiments, fluorochrome-conjugated secondary antibodies that have been cross-absorbed against multiple species are recommended to avoid cross-reactions.

Protocol 8

Immunostaining for proteins

Equipment and reagents

- Coplin jar
- Incubator at 37 °C
- Primary antibody against protein of interest
- 1% BSA in PBS
- PBS
- Fluorescence-conjugated secondary antibody
- 0.1% Triton X-100 in PBS
- 4% formaldehyde in PBS

Method

1 Equilibrate coverslip in PBS.

Protocol 8 continued

2 Dilute primary antibody in PBS, 1% BSA and apply 20 μl to coverslip.

3 Incubate for at least 1 h at 37 °C.

4 Wash at 25 °C for 10 min each in:

(a) PBS.

(b) PBS, 0.1% Triton X-100.

(c) PBS.

5 Dilute fluorescent-conjugated secondary antibody in PBS, 1% BSA to a concentration of approx. 3 μg/ml and apply 20 μl to coverslip.

6 Incubate for at least 20 min–1 h at 37 °C.

7 Repeat washes in step 4.

8 Fix signal in 4% paraformaldehyde, PBS for 10 min at 25 °C.[a]

9 Wash twice with PBS, 5 min each.[a]

10 Proceed with *in situ* hybridization protocol.[a]

[a] When the experiment calls for hybridization after immunodetection of specific proteins.

6.3 Chromosome paints and RNA FISH

Chromosome paints have proven useful for identifying individual chromosomes in cytogenetic preparations of metaphase cells and in delineating the boundaries of chromosome territories in interphase nuclei. An interesting example of this application is the study of the XIST RNA, which initiates inactivation of the X chromosome in mammalian nuclei (7). The chromosome paint when coupled with RNA hybridization to XIST delineates which X chromosome is inactivated and to what extent XIST RNA coats that chromosome (*Figure 6*; see also *Plate 12*). As with other sequential hybridizations, RNA species are detected first and then

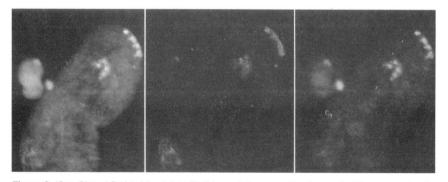

Figure 6 (See *Plate 12*) Hybridization to RNA in nuclei was key for discovering the novel role of XIST RNA, which 'paints' its parent X chromosome and induces its inactivation. Comparison of human XIST RNA to X chromosomal DNA; the latter was detected using a library of X chromosome-specific DNA fragments. In the cell line depicted (karyotype 47, XXX), two of the X chromosomes (red) are inactivated and coated by XIST RNA (green), whereas the active X chromosome (lower left) is not (7).

the chromosomal DNA is hybridized (see above). In some cases in which a chromosome paint is weak or not fully delineating the chromosome, we have obtained a more complete hybridization using cells prepared according to Kurz *et al.* (8) (see also Chapter 7). Not all commercial sources of chromosome paints work equally well in our hands. We have had success with paints that are prepared by degenerate PCR amplification of FAC-sorted chromosomes, such as CamBio's STAR*FISH paints. The following is a protocol for DNA hybridization using a chromosome paint as a probe, which can be coupled to RNA detection as mentioned above.

Protocol 9

Hybridization with chromosome paint

Equipment and reagents

- See *Protocol 5*
- 37 °C water-bath
- 70 °C heat block
- 20 μg Cot-1 DNA
- STAR*FISH paint (Cambio/Vysis)

Method

1 Heat denature the cellular DNA according to *Protocol 5*, steps 1–5.

2 Lyophilize 20 μg of Cot-1 and resuspend in 15 μl of commercially labelled STAR*FISH paint pre-warmed in a 37 °C water-bath for 30 min.

3 Heat denature the probe at 70 °C for 10 min.

4 Reanneal with the provided competitor DNA at 37 °C for 20 min to minimize non-specific hybridization.

5 Apply to coverslip and continue with *Protocol 5*, step 6.

6.4 Differentiating transcripts with intron and cDNA probes

Visualizing RNA detected with a cDNA probe in one colour and a probe specific for intron sequences in another can reveal much about the splicing of a transcript. When combined with immunolocalization of splicing factors, this can demonstrate structural information about where and how this process occurs (19). For example, Johnson *et al.* (21) used this strategy to compare the fate of RNAs produced from the normal and the splice-defective allele of collagen Iα1 in heterozygous human cells from a patient with osteogenesis imperfecta (*Figure 7*; see also *Plate 13*). Such analyses can demonstrate whether the mutant transcript is degraded or retained within the nucleus, and provide a means to better understand where and why the splice-defective transcript may fail to be transported to the cytoplasm.

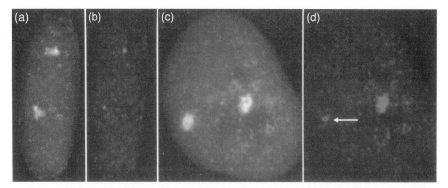

Figure 7 (See *Plate 13*) High resolution analyses of intron and exon RNA distributions can localize sites of splicing and detect splice defects from an individual allele in a heterozygous cell. Simultaneous RNA hybridizations with probes for COL1A1 intron 26 and cDNA in normal WI-38 fibroblasts (A, B) versus 054 mutant cells (C, D). The mutant cell line is heterozygous for a mutation in the splicing of intron 26. Both alleles in the normal cell show a limited overlap and polar orientation of intron 26 (red) with respect to the cDNA accumulation (green). In contrast, in the mutant cell line, transcripts from the mutant allele are readily discerned from the normal (arrow) by the more uniform distribution of intron 26 (red) throughout the cDNA track (green) (21).

We have successfully hybridized to introns using oligonucleotide probes (see above) and specific intron sequences amplified from genomic DNA by PCR (*Figure 11*; see also *Plate 16*). The intron of interest is amplified following the instructions provided in the PCR core kit (Roche) and then purified by phenol/chloroform extraction from a low melt agarose gel. The DNA is nick translated according to *Protocol 3* and used for hybridization as detailed in *Protocol 4*. To verify specificity, we typically hybridize the PCR probe simultaneously with a cDNA probe containing a different label (*Figure 7*; see also *Plate 13*).

6.5 Exon suppression hybridization: an example of the use of specific competition

In order to visualize the signal generated by essentially all introns in a transcript, we devised a strategy that blocks detection of exons by hybridizing with a genomic DNA probe in the presence of excess, unlabelled cDNA competitor. This 'exon suppression' strategy (21) makes it feasible to detect with a single probe the collective distribution of introns, and it illustrates a broader concept for how competition can enhance the capabilities of the FISH approach. The strategy is outlined in *Figure 8* and illustrated in *Figure 9* (see also *Plate 14*). We have used sequential hybridizations, first with a labelled COL1A1 cDNA followed by exon suppression hybridization, to dissect the COL1A1 RNA transcript track (21). This analysis has shown that the bulk of the 51 introns in transcripts from the COL1A1 gene are removed at one end of the transcript accumulation, at or near the gene.

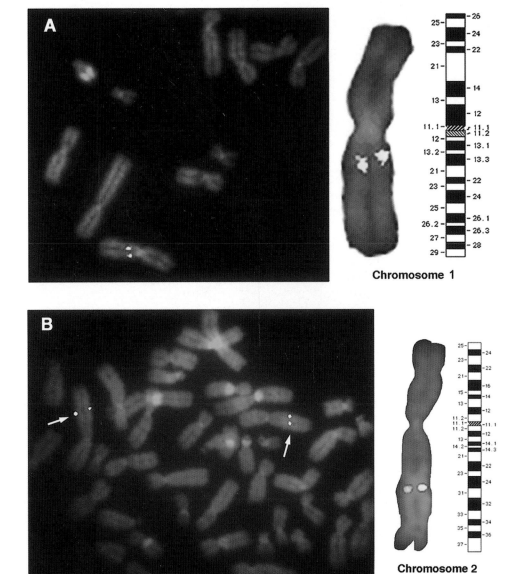

Plate 1 FISH mapping of single copy probes. Biotinylated DNA probes hybridized to normal human metaphase chromosomes counterstained with DAPI was detected with FITC (see Protocols 8–10). The band assignment was determined by band pattern analysis and fractional chromosome length measurement. (A) Doublet signals of a cDNA probe (1.7kb) in a partial metaphase localized to lq13.3 (9). (B) Doublet signals of a genomic PAC probe (~100kb) mapped to 2q31 on both homologues.

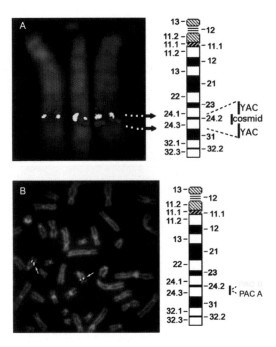

Plate 2 Relational mapping on chromosome 14 using two fluorochromes. (A) Three probes: Pairwise mapping of two YACs and one cosmid within a YAC contig (14q24.2–24.3). Two YACs, and one cosmid were detected with either rhodamine (digoxigenin, red) or FITC (biotin; yellow). Left chromosome: YAC1 (red) cosmid (yellow); middle chromosome: YAC2 (red) plus cosmid (yellow); right chromosome: YAC1 (yellow) plus YAC2 (red). The order shown is cen–YAC 1–cosmid–YAC2. The double hybridization signals often noted with YAC2 (right chromosome) suggest two domains of strong hybridization or Alu sequence clustering. (B) Two probes:DIG-labelled PACA (red) maps telomeric to biotin-labelled PACB, (yellow) at 14q24-31.

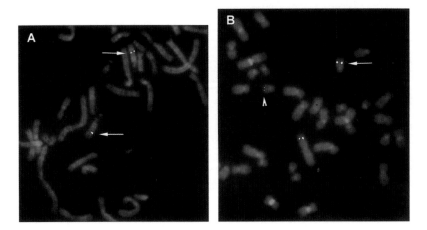

Plate 3 Mapping across a chromosome breakpoint. (A) Cosmid probe (14 kb) showing doublet hybridization signals at 14q24 on both chromosome homologues. (B) In a patient with a 2:14 (p25;q24) translocation, the same probe demonstrates doublet signals on both normal (arrow) and derivative (arrowhead) chr 14, as well as doublet signals on the short arm of derivative chr 2 (no arrow). A consistently stronger signal on derivative chr 2 suggests breakpoint is located toward the proximal (centromeric) end of the cosmid.

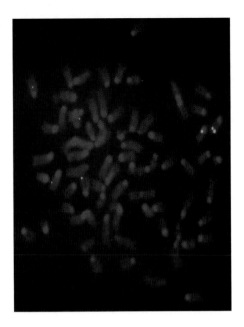

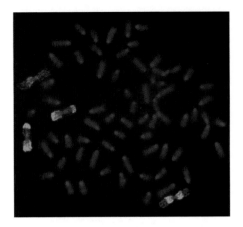

Plate 5 Painting of primary plasmacytoma metaphase spreads with paints for chromosomes 6 (red) and 15 (green). Note the reciprocal translocation (rcpT Rb6;15). For details, see Protocol 15.

Plate 4 Mouse metaphase chromosomes hybridized with DHFR (19) that was labelled by digoxigenin and visualized by an anti-digoxigenin–FITC antibody. For details, see Protocols 13 and 14.

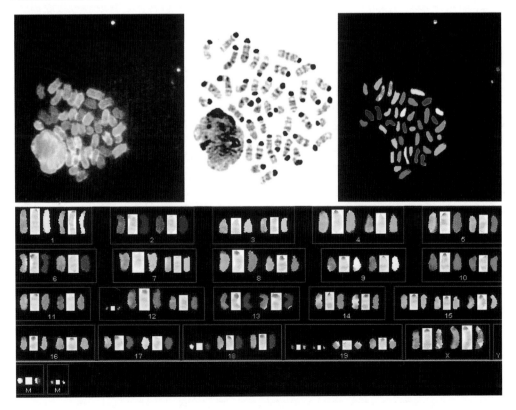

Plate 6 Banding of mouse chromosomes after SKY. For details see Protocols 16 and 17 and ref. 21.

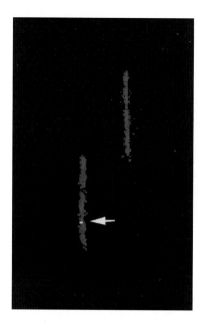

Plate 7 Fibre FISH analysis.
The FISH signal (arrow) shows the
location of a gene of interest
on a typical fibre FISH preparation.

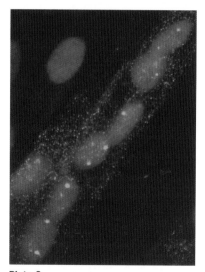

Plate 8

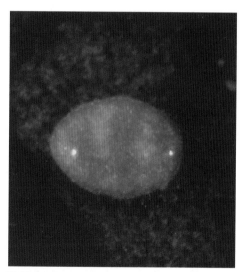

Plate 9

Plate 8 Detection of high and low abundance transcripts in nuclei versus cytoplasm of skeletal muscle
fibers. Male human multinucleated myotube nuclei show two muscle-specific myosin heavy chain (MyHC)
RNA foci (green) and a single dystrophin (X chromosome) RNA signal (red). MyHC mRNA is evident in the
myotube cytoplasm (green), whereas dystrophin RNA is detected only as focal accumulation at the gene in
the nucleus. A single undifferentiated myoblast nucleus (upper left) expresses neither MyHC nor dystrophin
RNA (1).

Plate 9 Sequential hybridizations to collagen type I α1 (COL1A1) RNA (red), followed by the COL 1A1 gene
(green) in a human fibroblast reveals nuclear RNA foci emanating from the gene (overlap appears yellow),
and distributed in the cytoplasm. More dispered nuclear signal likely represents transcripts moving towards
the nuclear envelope (19).

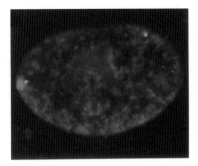

Plate 10 Nuclear RNA FISH can determine active alleles within a single cell. The UBE-1 gene escapes inactivation on the inactive X chromosome. The human XIST gene, expressed solely from the inactive X chromosome produces a large RNA accumulation (red) that "paints" the inactive X chromosome. In contrast, the much smaller UBE-1 nuclear RNA foci (green) are more typical of signal for protein coding genes (7).

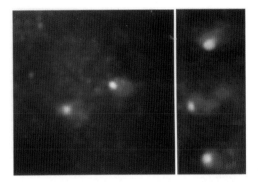

Plate 11 Simultaneous visualization of a gene and RNA emanating from that gene, shown for MyHC (right) and COL1A1 (left). Visualization of nuclear RNA (red) in direct spatial relation to the gene (green) using a dual band-pass filter (19), can determine allele activity, and identify structural details with potentially important biological implications. The COL1A1/MyHC RNA "tracks" (red) are both coincident and larger than the DNA signal (may extend 0.5-3μm beyond the gene) and consistently positioned to one side of it (1,19,21).

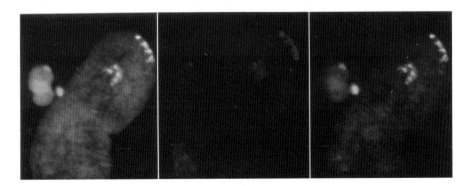

Plate 12 Hybridization to RNA in nuclei was key for discovering the novel role of XIST RNA, which "paints" its parent X-chromosome and induces its inactivation. In the cell line depicted, (karyotype 47, XXX), two of the X chromosomes (red) are inactivated and coated by XIST RNA (green), whereas the active X chromosome (lower left) is not (7).

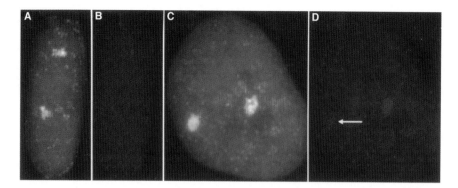

Plate 13 High-resolution analyses of intron and exon RNA distributions can localize sites of splicing and detect splice defects from an individual allele in a heterozygous cell. Simultaneous RNA hybridizations with probes for COL1A1 intron 26 and cDNA in normal WI-38 fibroblasts (A, B) versus 054 mutant cells heterozygous for a splicing mutation of intron 26 (C, D). Both alleles in the normal cell show a limited overlap and polar orientation of intron 26 (red) with respect to the cDNA (green). In the mutant cell, mutant allele transcripts are readily discerned from the normal (arrow) by more uniform distribution of intron 26 (red) throughout the cDNA track (green) (21).

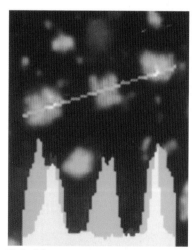

Plate 14

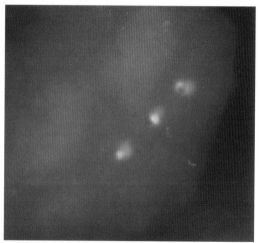

Plate 15

Plate 14 Selective competition coupled with high-resolution quantitative microscopy can enhance the power of FISH analyses. An "exon suppression" strategy (see Chapter 6, Figure 8) was devised to determine the collective distribution of fifty different COL1A1 intron RNAs (21). (Top) Collective intron RNA signal (red) was directly compared to spliceosome assembly factor SC-35 signal (green), demonstrating removal of most or all introns in the periphery of the SC-35 domain. (Botton) Microfluorimetry allows high-resolution quantitation of the relative distributions of these two fluorescent signals. Signal intensity readings of each pixel along the blue line shows quantitative removal of introns (splicing) at the periphery of the SC-35 domain. RNA detected with cDNA probe distributes throughout the domain (not shown) indicating spliced transcripts enter domain after splicing (21).

Plate 15 Triple-label detection of DNA, RNA, and protein. HeLa cell nucleus Cajal (coiled) bodies stained with an anti-coilin antibody (blue), coupled with sequential RNA/DNA hybridization of U2 RNA (red) and DNA (green), using a 6.1 kb genomic U2 locus probe. Nucleus shows a Cajal body with each of the three U2 gene foci, and a separate RNA signal (20).

Plate 16

Plate 17
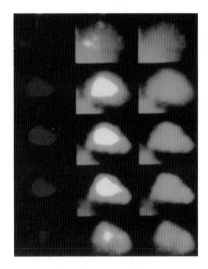

Plate 16 Spatial distribution of two different introns reflects the temporal order of COL1A1 transcript splicing. Simultaneous detection of COL1A1 RNA intron 24 (green) using overlapping biotinylated oligonucleotides, with a digoxygenin-labelled (nick-translation) PCR-amplified fragment of intron 26 (red), showed intron 24 (green) is spliced later, as it is retained throughout much of the globular RNA "track". Intron 26 (red) is only present in a small focus at one end of the track near the gene (see ref 21).

Plate 17 3-D information obtained by image capture of focal planes along the z-axis. A simplified 3-D analysis demonstrates collagen RNA (red) within an SC-35 domain (green), rather than just above or around it. Manual optical sectioning shows the two signals spanning several optical sections (~0.2 microns apart) going in and out of focus in essentially the same focal planes, and clearly overlapping (yellow) within the inner planes of the SC-35 domain (19). For higher resolution, a more sophisticated analysis would be useful (Fig 13; also Plate 18).

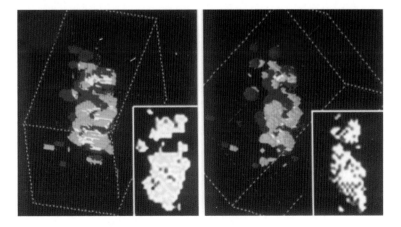

Plate 18 Deconvolution and 3-D rendering of XIST RNA distributes throughout interphase X-chromosome territory, not just above or below it. Optical sections captured with a CCD camera were deconvolved to remove out of focus light, and data rendered to display the 3-D object. The two panels depict a 90° rotation in the projected 3-D reconstruction (7). Each panel shows a surface view of the 3-D object, showing XIST RNA (red) and X chromosome DNA (green), with areas of overlap (white). The insets show similar 3-D objects are seen when only overlapping pixels are viewed.

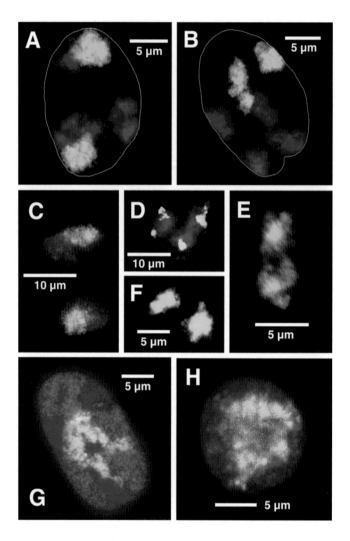

Plate 19 (A, B) – Projections of confocal series of sections through human diploid fibroblast nuclei with painted chromosome territories; each chromosome paint was labelled with a different hapten and detected with one fluorochrome (c.f. Table 2). Nuclear borders are outlined with a thin white (A) Green: chr 5,digoxigenin, FITC, red: chr 11, biotin, Cy 5; blue: chr 15, estradiol, Cy 3. (B) Green: chr 3, DNP, FITC; red: chr 11, biotin, Cy 5; blue: chr 5, digoxigenin, Cy 3. (C) Chromosome 1 painted with arm specific probes: 1p (red) and 1q (green); projection of a series of optical sections through nucleus from a human primary fibroblast. (D) Two X chromosomes (red) painted in a human female fibroblast, X active (right), X inactive (left), with bands p22.1-22.3 and q25-28 (green) labeled by hybridization with microdissected band-specific probes. (E) Extended focus image of a confocal series of sections through two painted chromosome 17 territories (green) with stained centromere region (red) in a human lymphocyte. (F) Two chromosomes 15 (green) with detected Prader-Willi syndrome regions (red) in a human fibroblast. (G) Projection of confocal series of sections through the nucleus of a human primary fibroblast; two pools of chromosome paints, chromosomes 1-5 and X (red) and chromosomes 17-20 (green) were hybridized; nuclear chromotain counterstained with propidium iodide (blue). Note small chromosomes more centrally andlarge chromosomes more peripherally positioned in the flat fibroblast nucleus. (H) Confocal optical section through a nucleus of a chicken neuron. Three pools of chromosome paints were hybridized: macrochromosomes (red), microchromosomes (green) and chromosomes of medium size (blue); note microchromosomes positioned centrally compared to macro- and medium-sized chromosomes.

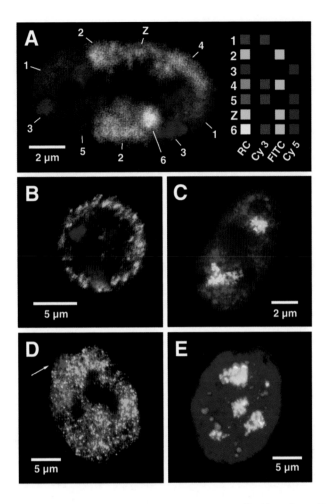

Plate 20 (A) Confocal section through the nucleus of a chicken primary fibroblast after multicolour FISH with combinatorially labelled painted probes (biotin, digozigenin, and estradiol) for seven macrochromosomes (1-6 and Z); detected using FITC, Cy 3, and Cy 5; scheme on right shows resulting colours (RC) and fluorochrome combinations used for each chromosome. (B) Confocal optical section through a lymphocytenucleus with painted chromosomes (18, red and 19, blue) and replication labelled chromatin (green)revealed after BrdU detection (see Protocol 11, A). Note peripheral labeling pattern characteristic of the mid-late S-phase with only chromosome 18 containing a proportion of late-replicating chromatin. (C) Murine(C2C12 myoblast) nucleus in early S phase showing chromosome 3 territories (green) and GFP-PCNA protein(red). GFP-PCNA is stably expressed as an N-terminal GFP fusion construct in these cells (kindly provided byDr M. C. Cardoso). Mouse-anti-GFP and sheep-anti-mouse-Cy3 antibodies (see Table 5, right column) wereused to refresh GFP-fluorescence lost during pretreatments for 3D FISH. (D) Nucleus of a human femalefibroblast with simultaneously visualised chromosome X territories (red) and hyperacetylated histone 2B(green). Histone was detected using rabbit-anti-hyperacetylated histone 2B (kindly donated by Dr B. M.Turner) and biotinylated goat-anti-rabbit. Digoxigenin of X chromosome paint and biotin of the secondaryantibodies were detected simultaneously after hybridization (see Table 5, middle column). Note somepatches of chromatin on nucleus periphery and X inactive chromosome territory (Barr body, marked by arrow) do not show hyperacetylated histone: these regions are transcriptionally inactive, nucleoli are not staineddue to small amount of chromatin. (E) Projection of confocal series of sections through a human fibroblastafter FISH with pan-centromeric probe (red) with subsequent detection of Ki-67 proteins (green), andpropidium iodide counterstaining (blue). Ki-67 antigens are not altered during 3D FISH pretreatments, andcan be detected after FISH (see Table 5, left column).

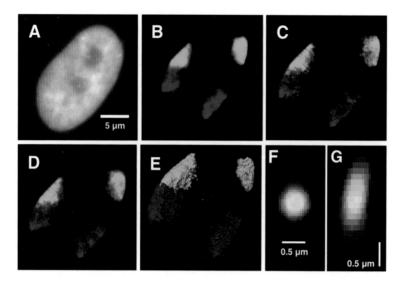

Plate 21 (A-E) Nucleus of a human diploid fibroblast with chromosomes 2 (red) and 3 (green) territories revealed by 3D FISH with chromosome specific paint probes (digoxigenin + Cy 3 and biotin + FITC, respectively). (A, B) Images collected using an epifluorescence microscope and a CCD camera: counterstaining with DAPI (A) and (B) chromosome territories. (C) A section chosen from the middle of aseries of confocal sections; note complex shape of the chromosome territories and uneven chromatin densityacross the territory. (D) Maximum intensity projection of the confocal series. (E) Surface rendering of theimage stack done with Amira (TGS, Merignac, France). (F,G) One Tetra Speck 500 bead recorded in twocolours on a Leica TCS SP. Green: 488 nm excitation; Red: 514 nm excitation. Xy(F) and xz(G) section.

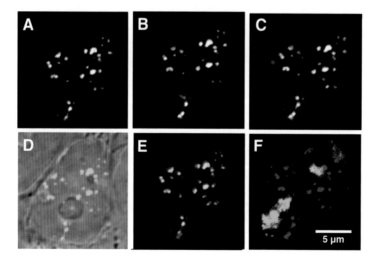

Plate 22 Preservation of the chromatin structure in neuroblastoma cell nucleus after fixation and 3D FISH. (A-C) – Optical sections from three different stacks of confocal serial sections through a late S phase nucleus of the same cell after replication pulse labelling using Cy3-dUTP microinjection. Matching optical sections from the nucleus in the same living cell (A), cell after fixation (B), cell after 3D FISH (C). Note the similar pattern of the replication foci, and the same shape and size on all three sections. (D) Overlay of a phase contrast image of the nucleus in the living cell and replication labelling (green) from the optical section shown on (A). (E) Overlay of optical sections from living cell nucleus (A) (green) and the same nucleus after 3D FISH (C) (red). Overlay shows replication-labelled chromatin has not significantly moved or dispersed after fixation and hybridation. (F) The same optical section as in (C) with both replication-labelled chromatin foci (red) and painted chromosome 5 territories (green).

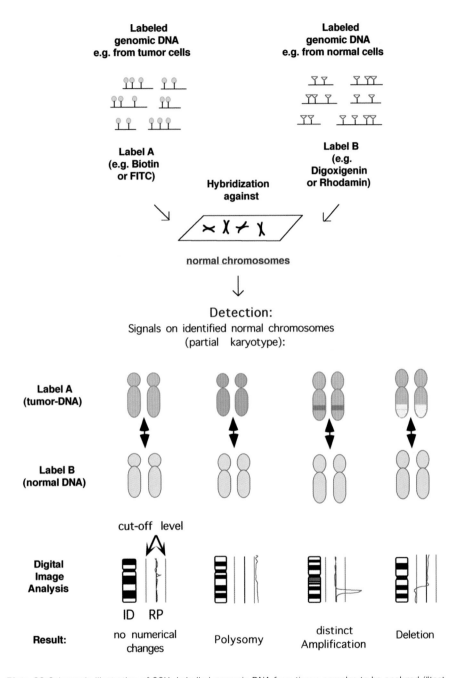

Labeled genomic DNA e.g. from tumor cells

Label A (e.g. Biotin or FITC)

Labeled genomic DNA e.g. from normal cells

Label B (e.g. Digoxigenin or Rhodamin)

Hybridization against

normal chromosomes

Detection:
Signals on identified normal chromosomes
(partial karyotype):

Label A (tumor-DNA)

Label B (normal DNA)

cut-off level

Digital Image Analysis

ID   RP

Result: no numerical changes   Polysomy   distinct Amplification   Deletion

Plate 23 Schematic illustration of CGH. Labelled genomic DNA from tissue samples to be analyzed ("test DNA") and differently labeled DNA from normal cells ("control DNA") are hybridized simultaneously to normal metaphase chromosomes. Chromosomal sequences present in extra copies in the test DNA (amplification or polysomies) result in a higher staining, and losses of chromosomal sequences (deletions or monosomies) result in a weaker staining respectively at the corresponding chromosomal target sequences as compared to the control DNA. Measurement of fluorescence intensities along metaphase chromosomes by digital image analysis results in intensity ratio profiles (RP), which are plotted along chromosome ideograms (ID). Over- or under-representations of the corresponding chromosomal regions are indicated by ratio profiles exceeding defined cut-off levels.

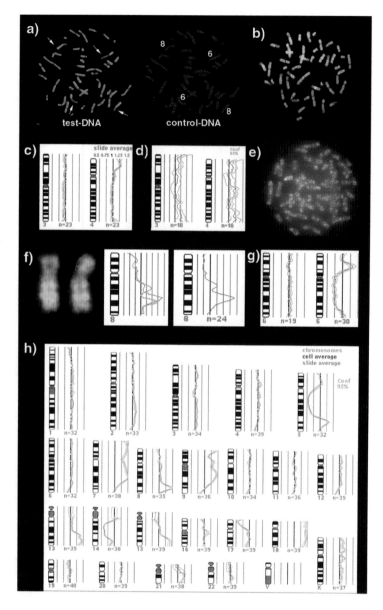

Plate 24 Examples of quality variation in CGH experiments. (a) CGH-experiment of good quality. A strong and homogeneous hybridization of the test- and control DNA is visible. Some imbalances are visible by eye (arrows). Note good chromosome separation. (b) Poor illumination (strong in the upper part, low at the bottom) of the microscopic field due to insufficient adjustment of the optical path. (c) Average ratio profiles of chromosomes 3 and 4 with a small 95% confidence interval indicating high quality CGH data. (d) Averageratio profiles of chromosomes 3 and 4 with a wide 95% confidence interval due to weak and granular hybridization. (e) Example of granular hybridization signals. (f) "Flattening" of CGH average ratio profiles. The individual ratio profiles (center panel) of two hybridized chromosomes 8 from the same case (left panel) are averaged and flattened (right panel). Before averaging the ratio profiles are adjusted to predefined lengths, resulting in a shift of the high level amplification ratio peaks. In extreme cases narrow amplifications and deletions remain undetected. (g) Average ratio profiles after CGH with DOP-PCR amplified DNA derived from 10 cells. Left: Unfixed cells resulted in correct CGH results, e.g. balanced state of chromosome 6. Right: Ethanol-acetic acid fixed cells from the same cell line give false-positive CGH data. (see text). (h) CGH ratio profiles using DNA from a single HL-60 cell (protocol 7). Note the narrow 95%-confidence intervals. In this example, no false-positive or false-negative results were obtained.

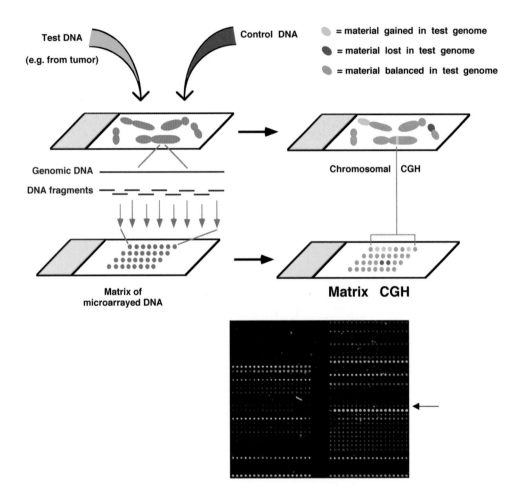

Plate 25 Schematic representation of the matrix-CGH approach. Chromosome targets are replaced by sets of well characterized genomic DNA fragments. Below the image of the chip where the target DNA spots are arrayed in replicas of twenty and CGH can detect highly amplified genes (see arrow).

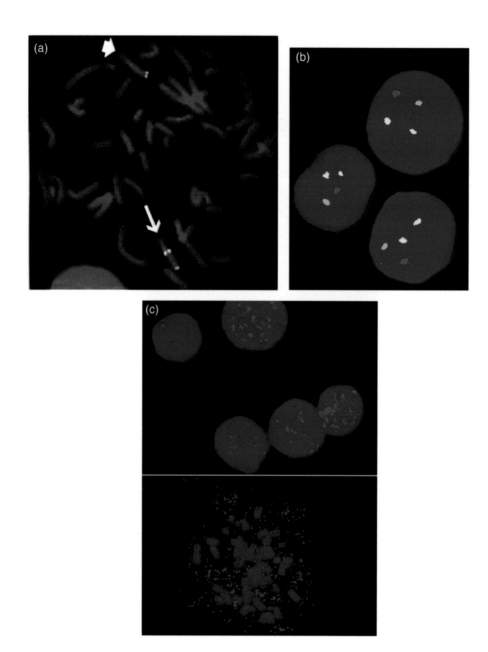

Plate 26 (a) Microdeletion analysis. The normal chromosome 7 (long arrow) shows yellow signals at both
the telomeric control region and the Williams syndrome region at 7q11.23. The abnormal chromosome
(short arrow) shows only the control signal, indicating a deletion at the Williams syndrome locus. (Photo
courtesy of Dr Gordon DeWald, Mayo Clinic. Probes from Oncor, Inc.) (b) D-FISH. The Philadelphia
chromosome is detected in interphase nuclei in leukaemia cells by the presence of two double fusion signals
(yellow), along with the BCR gene (red) and the ABL gene (green). (c) MYCN FISH. Top panel shows red
signals for the MYCN oncogene in a neuroblastoma. The cells show different levels of MYCN copy number
showing the heterogeneity of the tumour. The lower panel shows a metaphase from the same patient with
high levels of MYCN amplification in the form of double minutes.

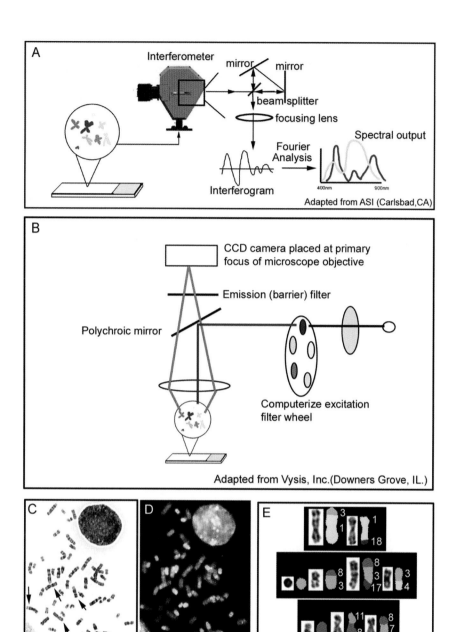

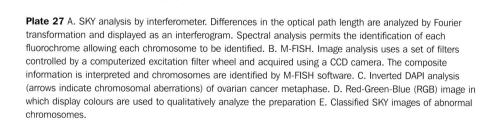

Plate 27 A. SKY analysis by interferometer. Differences in the optical path length are analyzed by Fourier transformation and displayed as an interferogram. Spectral analysis permits the identification of each fluorochrome allowing each chromosome to be identified. B. M-FISH. Image analysis uses a set of filters controlled by a computerized excitation filter wheel and acquired using a CCD camera. The composite information is interpreted and chromosomes are identified by M-FISH software. C. Inverted DAPI analysis (arrows indicate chromosomal aberrations) of ovarian cancer metaphase. D. Red-Green-Blue (RGB) image in which display colours are used to qualitatively analyze the preparation E. Classified SKY images of abnormal chromosomes.

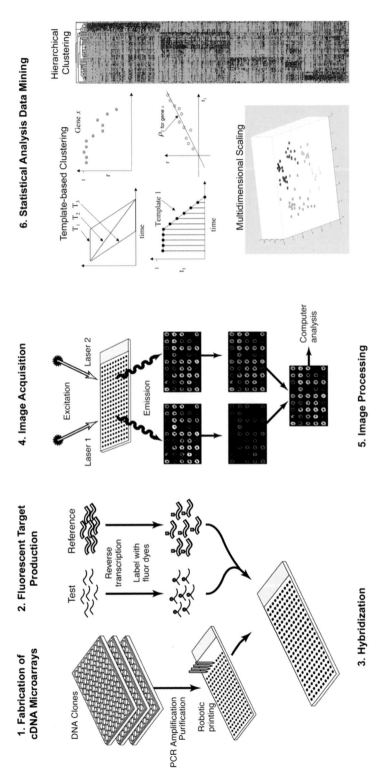

Plate 28 Overview of the fluorescent cDNA microarray hybridization procedure.

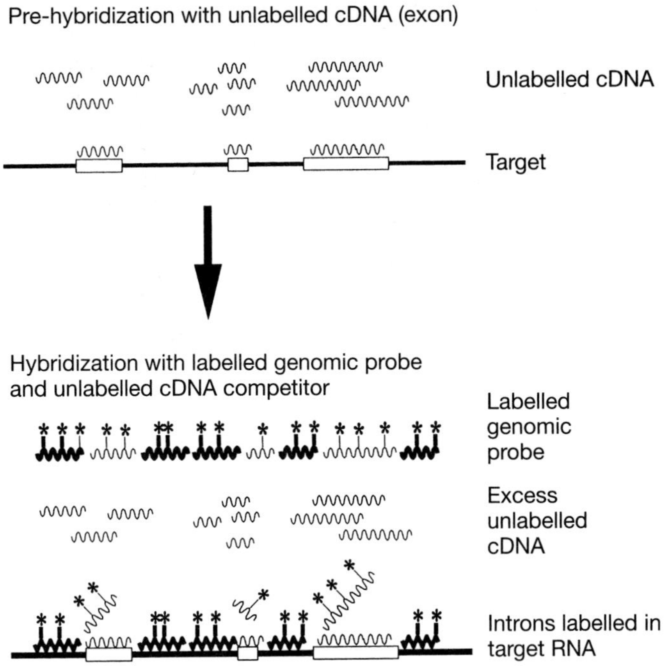

Pre-hybridization with unlabelled cDNA (exon)

Figure 8 Outline of the strategy for exon suppression hybridization. This technique was designed to visualize the collective distribution of all intron transcripts from a highly complex gene, by competing away any hybridization to exons by a labelled, full-length genomic DNA probe. Hybridization to exons is suppressed by a preliminary hybridization step with unlabelled, full-length cDNA, and addition of excess cold cDNA to the hybridization reaction (using a labelled genomic probe). Only labelled probe fragments containing intron sequences hybridize to their targets and are detected (solid bars).

This procedure is designed to localize the majority of the introns transcribed in a complex gene such as collagen. We note that the exon suppression technique would not necessarily detect a deviation in the level or distribution of any one intron, because a single intron may represent too small a proportion of a large, complex primary transcript. Also, it is important to assess the efficiency of suppression using a control experiment in which labelled, full-length cDNA is competed by unlabelled, full-length cDNA.

Protocol 10

Exon suppression

Equipment and reagents

- See *Protocols* 3 and 4
- Full-length cDNA
- Full-length genomic DNA
- DNase I (Roche)

Protocol 10 continued

Method

1   Digest >10 μg of full-length cDNA with DNase I to yield fragments approximately the same size as a nick translated probe.

2   A pre-hybridization step with 2–4 μg of the unlabelled, digested cDNA (step 1) is necessary to saturate the exon binding sites. 10 μg of each competitor, tRNA, Cot-1, and ssDNA are included in this pre-hybridization solution. Hybridize from 2 h to overnight at 37°C and wash without detecting.

3   In the second step, hybridize overnight with 50 ng nick translated probe that has been generated from a full-length genomic DNA clone. Include an additional 4 μg of the unlabelled, DNase I digested cDNA (step 1) and 10 μg each of tRNA, Cot-1, and ssDNA in this hybridization.

4   Wash and detect as in *Protocol 4*.

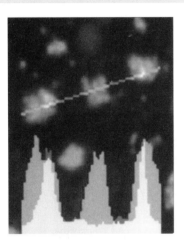

Figure 9 (See *Plate 14*) Use of selective competition procedures coupled with high resolution quantitative microscopy can enhance the power of FISH analyses. An 'exon suppression' strategy (diagrammed in *Figure 8*) was devised to determine the collective distribution of COL1A1 intron RNAs, encompassing 50 different introns (21). (Top) The collective intron RNA signal (red) was directly compared to the distribution of spliceosome assembly factor SC-35 (green), demonstrating that most if not all introns are removed in the periphery of the SC-35 domain. (Bottom) Microfluorimetry allows high resolution quantitation of the relative distributions of these two fluorescent signals. The computer reads out the signal intensity of each pixel along the blue line, showing quantitatively the removal of introns (splicing) occurs at the periphery of the SC-35 domain. Since RNA detected with a cDNA probe distribute throughout the domain (not shown), this indicates that spliced transcripts enter the domain after splicing (21).

7 Visualizing and analysing results

The hybridization protocols presented here can in principal be coupled with various kinds of detector reagents, for visualization by transmission microscopy (colorimetric detection), electron microscopy (gold or electron dense labels), or epifluorescence microscopy, as we describe here. Fluorescence microscopy provides a powerful combination of sensitivity, resolution, and multicolour labelling. Where the goal is simply to determine which cells within a culture or tissue are expressing a product, colorimetric detectors (e.g. alkaline phosphatase), may be as good or better than fluorescence because they avoid common difficulties with autofluorescence in tissues. However, such procedures do not provide adequate resolution for most studies of intracellular structure. Electron microscopy provides greater resolution than fluorescence, but requires thinner specimens, more sample preparation steps (potentially destructive to RNA), and thus far has proven less sensitive for detection of nucleic acids *in situ*.

FISH has been demonstrated to have robust sensitivity and resolution for the detection and localization of single genes and specific nuclear RNAs. It can routinely detect single genes or DNA sequences of 2–4 kb or more with high efficiency. As illustrated in the examples above, relatively low-level transcripts, such as intron-containing pre-mRNAs or pre-U2 RNA, can be readily detected with probes targeting as little as 22 bp; however, this sensitivity is likely augmented by small but localized concentrations of these transcripts. When RNA is dispersed throughout the cell or nucleus, one may see tiny fluorescent spots that likely represent the detection of individual mRNAs (~2–4 kb). However, this does not mean that the sensitivity of detection is a single transcript per cell. While it may be possible to visualize fluorescence from an individual transcript, unless there are many transcripts dispersed through the cell it is unlikely that one could confidently distinguish this from background. Hence, RNA distribution within the cell will impact substantially on the sensitivity of detection, since more localized signals are more easily discerned from background, as illustrated in *Figure 1* (see also *Plate 8*).

7.1 Microscopy

Because the confocal microscope can eliminate a significant proportion of fluorescent signal and can rapidly photobleach samples, we prefer to use a wide-field microscope for most purposes, even for most 3D analysis. (Confocal microscopy may be necessary for 3D studies of larger, brighter signals in thick tissue specimens; see Chapter 7.) High magnification, high numerical aperture lenses also greatly facilitate the detection of signals arising from single copy genes and low abundance RNA, and a microscope equipped with a ×100, 1.4 NA oil immersion objective is recommended. The size of most gene signals is at or below the level of resolution provided by a ×100 objective (approx. 250 nm) and, therefore, sequences up to 50–100 kb (or more) in length will produce a signal that appears as a single, small point of light. Because the limits of resolution are large relative to the size of individual molecules, this means that any consistent spatial separation evident between two signals has biological significance on a molecular scale *which is often overlooked* (*Figure 10*; see also *Plate 15*).

Filter selection also plays a critical role in fluorescence microscopy, particularly when precise alignment of signals is the experimental objective (see

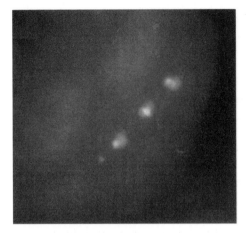

Figure 10 (See *Plate 15*) Triple-label detection of DNA, RNA, and protein. A HeLa cell nucleus stained with an anti-coilin antibody to visualize Cajal (coiled) bodies (blue), coupled with sequential RNA–DNA hybridization and two-colour detection of U2 RNA (red) and DNA (green), using a 6.1 kb genomic U2 locus probe. This nucleus shows a Cajal body with each of the three U2 gene foci, as well as a separate RNA signal (20).

Chapter 2). A triple bandpass filter set that employs a single beam splitter and emission filter incorporated into the body of the microscope (available from Chroma Technology) guarantees precise superimposition of signals from different excitation wavelengths. All microscope systems should be evaluated for optical shifts. A shift that is optical and not biological will result in one colour being consistently shifted in the same direction relative to the other.

Hundreds of cells can easily be viewed directly through the microscope by independent examiners, and in multiple experiments, which avoids biases that can be introduced by more selective analysis of a few imaged cells. Precise structural analysis can generate details concerning both the relative positions and morphology of two or more nuclear constituents which can lead to significant insights into nuclear processes (*Figures 2* and *5*; see also *Plates 9* and *11*).

7.2 Digital imaging

In recent years, the recording and analysis of microscopy data has been greatly aided by the development of CCD cameras specifically adapted for use with the light microscope. Generating digital images with CCD cameras offers several advantages over conventional photography, including the ease and speed of acquisition and storage, flexible manipulation of image size and contrast, signal morphometry and quantification, and the ability to capture and render 3D data sets.

While digital imaging can speed and enhance data analysis, care must be taken to record data in a manner that accurately reflects the signals in the actual sample. For example, because CCD cameras offer high sensitivity and digital images are easily manipulated, artefactual data can be generated by signals bright enough to 'bleed-through' the emission filter (e.g. bright TRITC signals through the FITC emission filter), background in the sample, or simply electronic noise, which can be misinterpreted as real signal. Conversely, very weak signal can be integrated over long exposures enabling visualization of signal not seen through the microscope. Digital imaging is an especially powerful technique because it can be used to quantify morphometric and intensity data. A 12-bit CCD camera has greater dynamic range than the human eye and, signal properties can be readily quantitated. However, results can be greatly affected by the way the measurement threshold is set. Microfluorimetric data is most meaningful when determined relative to a standard signal within the same image. As illustrated in a recent study where it was shown that a significantly increased accumulation of COL1A1 RNA was present at the mutant versus normal COL1A1 allele in osteogenesis imperfecta (*Figure 7*; see also *Plate 13*). Microfluorimetry can also be used to indicate the relative distributions of two different labels in the same nucleus, as shown for the analysis of spliced and unspliced COL1A1 nuclear structure (21) transcripts within a splicing factor enriched nuclear compartment, the SC-35 domain (*Figure 9*; see also *Plate 14*).

Although substantial 3D information can be gleaned by simply adjusting the focus while looking through the microscope eyepiece, more extensive analysis is sometimes important. 3D images can be generated by collecting *z*-series stacks of images from a wide-field microscope and applying a photon reassignment algorithm to reconstruct the image in 3D (for details see *Figures 12* and *13*, and *Plates 17* and *18*). Alternatively, a confocal microscope can be used to acquire 3D information, as discussed further in Chapter 7.

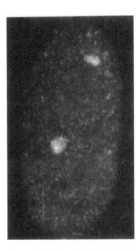

Figure 11 (See *Plate 16*) The spatial distribution of two different introns reflects the temporal order of splicing of COL1A1 transcripts. Detection of COL1A1 RNA intron 24 (green) using overlapping biotin labelled oligonucleotides simultaneously with intron 26 (red) using a nick translated probe (labelled with digoxigenin) from a PCR-amplified fragment (21). Because intron 24 (green) is spliced later, it is retained throughout much of the globular RNA 'track', whereas intron 26 (red) is only present in a small focus at one end of the track (near the gene) (see ref. 21).

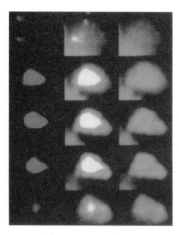

Figure 12 (See *Plate 17*) Valuable 3D information can be obtained simply by capturing images of distinct focal planes along the *z*-axis. In this example, a simplified approach to 3D analysis was used to make the point that collagen RNA (red) is found *within* an SC-35 domain (green), rather than just above or around it. By manual optical sectioning, the two signals, which both span several optical sections (~0.2 μm apart), can be seen going in- and out-of-focus in essentially the same focal planes, and clearly overlap (yellow) within the inner planes of the SC-35 domain (19). For higher resolution analysis, particularly for smaller structures, a more sophisticated 3D analysis would be more useful (*Figure 13*; see also *Plate 18*).

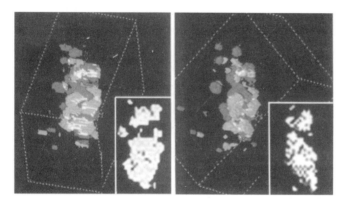

Figure 13 (See *Plate 18*) Deconvolution and 3D rendering shows that XIST RNA distributes throughout the interphase X chromosome territory, not just above or below it. Here optical sections captured with a CCD camera were deconvolved to remove out-of-focus light, and data from all planes then rendered to display the object in three-dimensions. The two panels depict a 90° rotation in the projected three-dimensional reconstruction (7). Each panel shows a surface view of the 3D object from that angle, showing XIST RNA (red) and X chromosome DNA (green), with areas of overlap (white). The insets show a similar 3D object is seen when only pixels with overlap are viewed.

8 Concluding remarks

This chapter provides a methodological overview and detailed experimental protocols, for the study of gene expression and RNA metabolism, as it relates to nuclear structure and genome/chromosome organization in the context of individual cells. Molecular cytology is invaluable for investigating not only cell type-specific gene expression during development, but also how complex molecular and biochemical processes of gene expression are actually integrated and working within the cell structure. For example, the *in situ* visualization of stable, spliced XIST RNA was essential for revealing the unprecedented 'painting' of the X chromosome by this non-coding RNA, key to its role in X inactivation. Other examples of the singular abilities of *in situ* analysis are the ability to visually discriminate differential transcription or splicing from alleles within a single nucleus, or address where a block to transport of mutant transcripts occurs within that nucleus. It is increasingly appreciated that the intracellular or intranuclear localization of genes, RNAs, and proteins *in situ* may be critically important for understanding their roles both in normal cells and disease.

With the continued development of new hybridization strategies, new types of probes, and innovative developments in microscopy for fixed and living cells, the power and versatility of this approach will be increasingly appreciated and essential for understanding fundamental aspects of cell function intimately intertwined with structure.

Acknowledgements

We thank Meg Byron and Kelly Smith for helpful discussions and critical reading of the manuscript. This work was supported by National Institutes of Health grants GM49254 and GM53234 to J. B. Lawrence and GM18846 to L. S. Shopland.

References

1. Smith, K. P., Moen, P. T., Wydner, K. L., Coleman, J. R., and Lawrence, J. B. (1999). *J. Cell Biol.*, **144**, 617.
2. Johnson, C. V. and Lawrence, J. B. (1991). *Genet. Anal. Tech. Appl.*, **8**, 75.
3. McNeil, J. A., Johnson, C. V., Carter, K. C., Singer, R. H., and Lawrence, J. B. (1991). *Genet. Anal. Tech. Appl.*, **8**, 41.
4. Lawrence, J. B. (1990). In *Genome analysis Volume 1: Genetic and physical mapping* (ed. K. E. Davies and S. M. Tilghman), p. 1. Cold Spring Harbor Press, Cold Spring Harbor NY.
5. Carter, K. C., Taneja, K. L., and Lawrence, J. B. (1991). *J. Cell Biol.*, **115**, 1191.
6. Xing, Y. and Lawrence, J. B. (1991). *J. Cell Biol.*, **112**, 1055.
7. Clemson, C. M., McNeil, J. A., Willard, H. F., and Lawrence, J. B. (1996). *J. Cell Biol.*, **132**, 259.
8. Kurz, A., Lampel, S., Nickolencko, J. E., Bradl, J., Benner, A., Zirbel, R. M., *et al.* (1996). *J. Cell Biol.*, **135**, 1195.
9. Lawrence, J. B., Singer, R. H., and Marselle, L. M. (1989). *Cell*, **57**, 493.
10. Lawrence, J. B., Marselle, L. M., Byron, K. S., Johnson, C. V., Sullivan, J. L., and Singer, R. (1990). *Proc. Natl. Acad. Sci. USA*, **87**, 5420.
11. Xing, Y., Johnson, C. V., Dobner, P. R., and Lawrence, J. B. (1993). *Science*, **259**, 1326.
12. Fey, E. G., Krochmalnic, G., and Penman, S. (1986). *J. Cell Biol.*, **102**, 1654.
13. Lawrence, J. B. and Singer, R. H. (1986). *Cell*, **45**, 407.
14. Verma, R. S. and Babu, A. (1989). *Human chromosomes: manual of basic techniques.* Pergamon Press, Elmsford, New York.
15. Gerdes, M. G., Carter, K. C., Moen, P. T., and Lawrence, J. B. (1994). *J. Cell Biol.*, **126**, 289.

16. Johnson, C. V., Singer, R. H., and Lawrence, J. B. (1991). *Methods Cell Biol.*, **35**, 73.

17. Lawrence, J. B., Villnave, C. A., and Singer, R. H. (1988). *Cell*, **52**, 51.

18. Matera, A. G., Frey, M. R., Margelot, K., and Wolin, S. L. (1995). *J. Cell Biol.*, **129**, 1181.

19. Xing, Y., Johnson, C. V., Moen, P. T., McNeil, J. A., and Lawrence, J. B. (1995). *J. Cell Biol.*, **131**, 1635.

20. Smith, K. P. and Lawrence, J. B. (2000). *Mol. Biol. Cell*, **11**, 2987.

21. Johnson, C. V., Primorac, D., McKinstry, M., McNeil, J. A., Rowe, D., and Lawrence, J. B. (2000). *J. Cell Biol.*, **150**, 417.

22. Carter, K. C., Bowman, D., Carrington, W., Fogerty, K., McNeil, J. A., Fay, F. S., *et al.* (1993). *Science*, **259**, 1330.

Chapter 7

FISH on three-dimensionally preserved nuclei

I. Solovei, J. Walter, M. Cremer, F. Habermann, L. Schermelleh, and T. Cremer
Human Genetics, Department of Biology II, Ludwig Maximilians University, Munich, Richard Wagner Str. 10/I, 80333 Muenchen, Germany.

1 Introduction

This chapter focuses on the fluorescence *in situ* hybridization (FISH) of DNA probes to three-dimensionally preserved cells, so-called 3D FISH. This technique allows three-dimensional visualization of specific DNA and RNA targets within the nucleus at all stages of the cell cycle. It provides information about the arrangement of chromosome territories and the organization of subchromosomal domains, about the pattern of chromatin density within a chromosome territory, about positions of individual genes and RNA transcripts read from them. Accumulation of such data is necessary for understanding relationships between the spatial organization of the genome and its functioning in the interphase nucleus.

Quantitative studies of the nuclear architecture using radioactive and non-radioactive *in situ* hybridization approaches began in the 1980s (1–13). The method is by now technically well developed (14). *Figures 1–3* (see also *Plates 19–21*) give some examples of 3D FISH on various nuclei. 3D FISH studies have demonstrated that chromosomes in the interphase nucleus occupy compact and mutually exclusive territories which are variable in shape (*Figure 1A, B*, and *Plate 19A,B*; *Figure 2A–D*, and *Plate 20A–D*) (15). Chromosome arms form separate domains within chromosome territories (*Figure 1C*; see also *Plate 19C*) (16) and the same holds true for medium sized (several Mbs) chromosomal subregions (*Figure 1D–F*; see also *Plate 19D–F*). Double minutes and homogeneously stained regions (HSRs) in the nuclei of neuroblastoma cells do not overlap with the normal chromosome territories and are located in the interchromosomal domain space between the territories (17).

The question as to what extent chromosome territories and chromosomal subregions maintain specific cell type- or cell cycle-dependent positions and spatial arrangement is still open. Some authors argue that chromosome arrange-

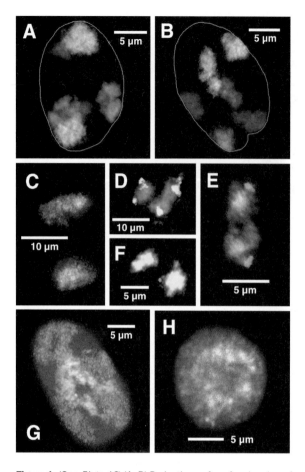

Figure 1 (See *Plate 19*) (A, B) Projections of confocal series of sections through nuclei of human diploid fibroblasts with painted chromosome territories; each chromosome paint was labelled with a different hapten and detected with one fluorochrome (compare *Table 2*). The borders of the nuclei are outlined with a thin white line (DAPI counterstaining, not detectable with the confocal microscope used, is not shown). (A) Green: chromosome 5, paint probe labelled with digoxigenin and detected with FITC; red: chromosome 11, biotin, Cy5; blue: chromosome 15, estradiol, Cy3. (B) Green: chromosome 3, paint probe labelled with DNP and detected with FITC; red: chromosome 11, biotin, Cy5; blue: chromosome 5, digoxigenin, Cy3. (C) Chromosome 1 painted with arm-specific probes: 1p (red) and 1q (green); projection of a series optical sections through nucleus from a human primary fibroblast. (D) Two X chromosomes (red) painted in a human female fibroblast, X active on right, X inactive on left, with bands p22.1-22.3 and q25-28 (green) labelled by hybridization with microdissected band-specific probes; image collected using a CCD camera and epifluorescence microscope. (E) Extended focus image of a confocal series of sections through two painted chromosome 17 territories (green) with stained centromere region (red) in a human lymphocyte. (F) Two chromosomes 15 (green) with detected Prader–Willi syndrome regions (red) in a human fibroblast; image collected using CCD camera and epifluorescence microscope. (G) Projection of confocal series of sections through the nucleus of a human primary fibroblast; two pools of chromosome paints were hybridized: for chromosomes 1–5 and X (red), and chromosomes 17–20 (green), nuclear chromatin counterstained with propidium iodide (blue). Note a more central position of small chromosomes and a more peripheral location of large chromosomes in the flat fibroblast nucleus. (H) Confocal optical section through a nucleus of a chicken neuron. Three pools of chromosome paints were hybridized: for macrochromosomes (red), for microchromosomes (green), and for chromosomes of medium size (blue); note a central position of microchromosomes in comparison to that of macro- and medium-sized chromosomes.

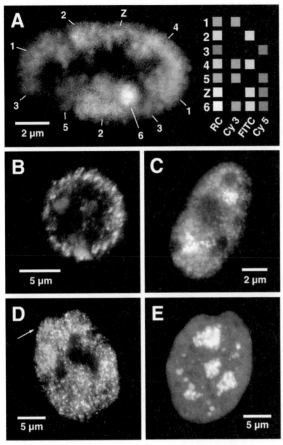

Figure 2 (See *Plate 20*) (A) Confocal section through the nucleus of a chicken primary fibroblast after multicolour FISH with paint probes for seven macrochromosomes (1–6 and Z); paints were combinatorially labelled with biotin, digoxigenin, and estradiol with subsequent detection using FITC, Cy3, and Cy5; the scheme on the right shows the resulting colours (RC) and fluorochrome combinations used for detection of each chromosome. (B) Confocal optical section through a lymphocyte nucleus with painted chromosomes (18, red and 19, blue) and replication labelled chromatin (green) revealed after BrdU detection (see *Protocol 11A*). Note a peripheral labelling pattern characteristic of the mid–late S phase; chromosome 18 contains a proportion of late replicating chromatin, while chromosome 19 does not. (C) Epifluorescence microscope image of a mouse cell (C2C12 myoblast) nucleus at the early S phase stage with visualized chromosome 3 territories (green) and GFP-PCNA protein (red). GFP-PCNA is stably expressed as N terminal GFP fusion construct in these cells (kindly provided by Dr M. C. Cardoso). PCNA pattern was initially visible due to GFP fluorescence, however this fluorescence was lost during pre-treatments for 3D FISH. To refresh the protein pattern mouse anti-GFP and sheep anti-mouse–Cy3 antibodies were applied (see *Table 5*, right column). (D) Epifluorescence microscope image of the nucleus of a human female fibroblast with simultaneously visualized chromosome X territories (red) and hyperacetylated histone 2B (green). Histone was detected using rabbit anti-hyperacetylated histone 2B (antibodies were kindly donated by Dr B. M. Turner) and goat anti-rabbit conjugated with biotin. Digoxigenin of X chromosome paint and biotin of the secondary antibodies were detected simultaneously after hybridization (see *Table 5*, middle column). Note that some patches of chromatin on the periphery of the nucleus and X inactive chromosome territory (Barr body, marked by arrow) do not show hyperacetylated histone: these regions are transcriptionally inactive; nucleoli are not stained due to a very small amount of chromatin. (E) Projection of the confocal series of sections through a human fibroblast after FISH with pan-centromeric probe (red) with subsequent detection of Ki-67 proteins (green), and propidium iodide counterstaining (blue). Ki-67 antigens do not alter during pre-treatments needed for 3D FISH and can be detected after FISH (see *Table 5*, left column).

ment in human interphase nuclei is random (18) while others support the possibility of a preferential positioning of some chromosomes (19, 20). Recently, Croft and co-workers (21) reported the preferential location of the gene-poor chromosome 18 territory in a more peripheral position, and the gene-rich

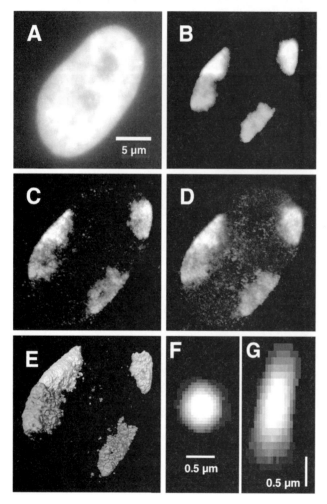

Figure 3 (See *Plate 21*) (A–E) Nucleus of a human diploid fibroblast with chromosomes 2 (red) and 3 (green) territories revealed by 3D FISH with chromosome-specific paint probes (digoxigenin + Cy3 and biotin + FITC, respectively). (A, B) Images collected using an epifluorescence microscope with a CCD camera: counterstaining with DAPI (A) and chromosome territories (B). (C) A section chosen from the middle of the stack of the series of confocal sections; note the complex shape of the chromosome territories and uneven chromatin density across the territory. (D) Maximum intensity projection of the confocal series. (E) Surface rendering of the image stack done with Amira (TGS, Mérignac, France). (F, G) One Tetra Speck 500 bead recorded in two colours on a Leica TCS SP. Green: excitation with 488 nm, detected emission from 495–540 nm. Red: excitation with 514 nm, detected emission from 570–645 nm. *xy-* (F) and *xz-* (G) section.

chromosome 19 in a more central position of the nuclei in cycling cells (*Figure 2B*; see also *Plate 20B*). We found that in cell nuclei from different species small chromosomes tend to locate more centrally than larger ones (*Figure 1G, H*; see also *Plate 19G, H*).

The equipment, as well as the protocols for probe labelling, hybridization, and detection required for 3D FISH, are basically the same as for FISH to metaphase chromosomes (2D FISH). Chapters 2 and 3 in this edition provide detailed protocols for probe labelling and 2D FISH. The principal requirement of 3D FISH is the preservation of native chromatin arrangements in fixed nuclei. Therefore this chapter focuses on protocols for 3D FISH emphasizing the methods of fixation and further treatment of fixed cells (Sections 2–5). These steps are crucial because two contradictory goals have to be combined in 3D FISH:

(a) Preserving the nuclear morphology as much as possible.

(b) Making nuclear DNA accessible for probe hybridization in order to obtain a good FISH signal.

Of equal importance for a quantitative 3D analysis of chromosomal structure is the use of confocal laser scanning microscopy (CLSM), image analysis, and 3D reconstruction (see Section 10).

2 Preparation and fixation of cells

Cells from different sources require appropriate modifications of fixation protocols. We supply three protocols:

(a) *Protocol 2* for adherent cells, e.g. fibroblasts or epithelial cells.

(b) *Protocol 3* for cells growing in suspension, e.g. EBV-transformed lymphoblastoid cells or other haematopoetic cell lines.

(c) *Protocol 4* for cells directly isolated from peripheral blood.

Prior to cell fixation, slides have to be prepared as described in *Protocol 1*.

2.1 Preparation of slides

For confocal laser scanning microscopy we strongly recommend using cells attached to coverslips with defined optical properties (*Figure 4*, see also Section 10.2). The set-up shown in *Figure 4* minimizes optical aberrations caused by image collection through the mounting medium and allows high resolution studies of intranuclear structures, as well as precise distance measurements (22). Depending on the intended methods of microscopy and image analysis, cells may also be cultured on conventional microscope slides with a thickness of about 1 mm. Microscope slides are more convenient to handle, however coverslips are preferable for CLSM when high quality 3D data are needed.

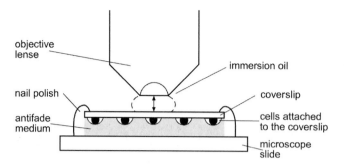

objective
lense

immersion oil

nail polish

coverslip

antifade
medium

cells attached
to the coverslip

microscope
slide

Figure 4 Mounted preparation after 3D FISH observed under microscope. For optimal performance, the cells should be directly grown on, or mounted on, the coverslip, to minimize optical aberrations that result from focusing through a thick layer of antifade medium.

In the following protocols we use the term 'slides' for both coverslips and microscope slides with attached cells. Frosted ends on conventional microscope slides allow marking with a pencil. Coverslips with frosted ends can be produced by treatment with hydrofluoric acid.

Protocol 1

Preparation of slides

Equipment and reagents

- Coverslip, 26 × 76 mm or smaller, thickness 0.17 ± 0.01 mm (Hecht Assistant)[a]
- QuadriPERM plus culture dishes (In Vitro Systems)
- Fine forceps
- Diluted hydrofluoric acid, 48%
- Ethanol/ether (1:1, v/v)

Method

1   Immerse one end of a coverslip into a small metal or plastic box that is filled up to 1 cm with hydrofluoric acid for 30 sec. Caution: hydrofluoric acid is highly aggressive!

2   Thoroughly rinse in ddH$_2$O.

3   Wash slides in a 1:1 mixture of ethanol/ether for at least 1 h.

4   Let slides air dry under sterile conditions and place them in sterile QuadriPERM chambers under sterile conditions until use.[b]

[a] These coverslips have an especially even thickness and an appropriate refractory index.

[b] This procedure is sufficient for sterilization; autoclaving of slides is not necessary. Flaming increases the danger of breaking slides (especially at the stage of freezing in liquid nitrogen) and should be avoided.

2.2 Cultivation and fixation of adherent cells

Substrate adhesive cells are routinely propagated in conventional plastic cell culture flasks. The cultures should be trypsinized during the logarithmic growth phase and seeded at appropriate density on slides. Most cell types can be successfully grown on glass slides, though they often grow slower than in plastic flasks.

Protocol 2

Subculture and fixation of adherent cells on slides

Equipment and reagents

- Sterile pre-treated slides in QuadriPERM dishes (see *Protocol 1*)
- CO_2 incubator
- Appropriate culture medium
- Phosphate-buffered saline (PBS): 140 mM NaCl, 2.7 mM KCl, 6.5 mM Na_2HPO_4, 1.5 mM KH_2PO_4 pH 7.2

- PBS with 0.04% Na azide
- Trypsin/EDTA: 0.05% (w/v) trypsin and 0.02% (w/v) EDTA in PBS; prepare from 10 × concentrated solution (Biochrom)
- 4% paraformaldehyde (PFA) (Merck) in PBS[a]

Method

1   Trypsinize cells from a 75 cm^2 culture flask under routine conditions and resuspend in 20–40 ml of medium. Transfer 5 ml of this cell suspension onto a slide placed in a QuadriPERM chamber (for most cell types one flask with an appropriate cell density[b] is sufficient for four to eight slides).

2   Allow cells to reach confluency.[b]

3   Wash slides with cells in PBS (37 °C), 2 × 3 min.

4   Transfer the slides to 4% PFA for 10 min at room temperature.

Note: from this point until mounting the preparation in antifade never let the cells dry out! We recommend that all steps such as washings, changing incubation media, etc. be performed by quickly transferring the slides from one Coplin jar to the next.

5   Wash in PBS, 3 × 5 min.

6   Proceed to pre-treatments described in *Protocol 5*.[c]

[a] Prepare fresh each time from powdered paraformaldehyde. Place the required amount of paraformaldehyde powder into a heat-resistant Duran bottle, add the corresponding amount of PBS, and stir at 60 °C until the solution becomes clear. Cool to room temperature before use.

[b] The duration of cell culture on slides depends on the starting and desired final cell density and their growth rate. For example, primary human fibroblasts should reach confluency on slides within three to five days.

[c] The cells can be kept in PBS containing 0.04% Na azide at 4 °C for several days.

2.3 Preparation, attachment, and fixation of cells growing in suspension

This section describes the handling of cells that grow in suspension, e.g. peripheral lymphocytes, EBV-transformed lymphoblastoid, or other haematopoetic cell lines. For fixation and subsequent 3D FISH, non-adherent cells must be attached to a slide under conditions that maintain their 3D morphology. For the attachment we recommend using poly-lysine. Cells growing in suspension are prone to shrinkage during fixation. We found that this can be prevented by incubating the cells in a high serum concentration (50%) during their attachment followed by a short incubation in a hypotonic buffer before fixation. The attachment of fixed cells by cytospin centrifugation as described in some publications (18, 21, 35) in our experience causes substantial flattening of the nuclei.

Protocol 3

Preparation and fixation of cells growing in suspension

Equipment and reagents

- Cell counting chamber
- Appropriate culture medium supplemented with 50% (v/v) fetal calf serum (FCS)
- Poly-lysine hydrobromide: M_r >150 000, 1 mg/ml (Sigma), make each time fresh from a 10 mg/ml stock solution in H_2O; stock can be kept at –20 °C

- PBS
- $0.3 \times$ PBS
- 4% paraformaldehyde (PFA) in $0.3 \times$ PBS (see *Protocol 2*)

Method

1 Cover an area of about 2×2 cm in the centre of a slide with 300 µl of a 1 mg/ml poly-lysine solution. Incubate for 1 h at room temperature, then drain poly-lysine off.

2 Briefly rinse slides in ddH$_2$O and air dry.[a]

3 Calculate the concentration of cells in the cell culture after careful resuspension by using a cell counting chamber.

4 Allowing approx. 1×10^6 cells/slide, transfer the required volume of cell suspension to a plastic tube, and centrifuge at 170 g for 10 min.[b]

5 Resuspend cells in 50% FCS/RPMI at a concentration of $\sim 3 \times 10^6$ cells/ml.

6 Transfer 0.3 ml of this cell suspension to the poly-lysine-coated area of each slide.[c]

7 Allow cells to attach at 37 °C in the incubator for 10–60 min. Monitor attachment of cells under the microscope; attaching lymphocytes and lymphoblastoid cells form characteristic cytoplasmic processes.

8 Drain off the medium and place the slide in $0.3 \times$ PBS for exactly 1 min.[d]

9 Fix in 4% PFA/$0.3 \times$ PBS for 10 min at room temperature.

Protocol 3 continued

10 Wash 3 × 5 min in PBS.

11 Proceed to pre-treatments described in *Protocol 5*. If needed, cells can be kept in PBS with 0.04% Na azide at 4°C for several days.

[a] Coating should be done directly or not more than few days before use.

[b] An excessive amount of cells is used as only a fraction of cells will attach to the poly-lysine-coated slide.

[c] At this stage, cell density on the first slide can be checked in a phase-contrast microscope and adjusted if necessary. We have found that the indicated number of cells usually results in a good separation of individual cells and a density that is convenient for microscopy after FISH.

[d] The exact duration of hypotonic treatment is important since longer incubation may affect the morphology of nuclei. Hypotonic treatment is known to cause swelling of nuclei which would alter the chromosome structure (34). The short hypotonic treatment with 0.3 × PBS (90 mOsm) followed by fixation in a solution of reduced osmolarity (4% PFA in 0.3 × PBS with approx. 220 mOsm) as suggested in the protocol retained the same nuclear diameter of approx. 9 μm after fixation as observed in the living cell (our own unpublished data).

3 Preparation of cells directly isolated from peripheral blood

Isolation of mononuclear blood cells, mainly lymphocytes, directly from peripheral blood is described in *Protocol 4*. The isolation technique is based on cell separation in a Ficoll gradient using LeucoSEP® tubes and is a modification of the procedure suggested by the manufacturer. To obtain cycling cells, peripheral T lymphocytes may be stimulated with phytohaemagglutinin (PHA). T lymphocytes start to enter S phase within 20–30 h after adding PHA to the culture medium (23). For non-stimulated cells, skip *Protocol 4*, steps 11–14. Various types of white blood cells can also be obtained by using magnetic beads coated with antibodies against surface antigens specific for certain cell types (24).

Protocol 4

Isolation of mononuclear cells directly from peripheral blood

Equipment and reagents

- 50 ml LeucoSep tubes (Greiner)
- 15 ml Falcon tubes
- Sterile plastic Pasteur pipette (3 ml)
- 25 cm² tissue culture flasks
- Appropriate culture medium
- Appropriate culture medium supplemented with 50% (v/v) fetal calf serum (FCS)
- Heparin-Na, 5000 U/ml (B. Braun)
- Ficoll–Paque™ PLUS (Pharmacia Biotech), kept at 4°C in the dark
- Phytohaemagglutinin (PHA) (Murex Biotech)

Method

1 Use fresh peripheral blood anticoagulated with heparin-Na (0.5–1 ml of blood/slide).

2 Warm Ficoll–Paque PLUS to room temperature.

3 Pipette 15 ml of Ficoll–Paque PLUS into a LeucoSEP tube and centrifuge at 1000 g for 30 sec. Ficoll–Paque is then under the filter disc of the tube.

4 Load up to a maximum of 25 ml of heparinized blood into the tube and centrifuge at 600 g for 15 min. Erythrocytes sink to the bottom of the tube, while the mononuclear cells form a visible layer between the plasma (on top) and the Ficoll layer, above the disc.

5 Aspirate the mononuclear cell layer with a Pasteur pipette (2–5 ml) and transfer into a 15 ml Falcon tube.

6 Add 10 ml of medium to the cell suspension and gently resuspend.

7 Spin down at 170 g for 10 min, and discard the supernatant.

8 Resuspend cells in 5 ml of medium by gentle aspiration with a Pasteur pipette.

9 Repeat steps 7 and 8.

10 If cells are not stimulated continue with *Protocol 3*, step 1 for attaching them on slides.

11 For the stimulation of T cells prepare 25 cm^2 culture flasks with 10 ml of pre-warmed (37 °C) medium in each.

12 Add the cell suspension to the culture flasks, allowing lymphocytes from 5–10 ml of whole blood per flask; add PHA to the medium to a final concentration of 4 μg/ml.

13 Incubate cells at 37 °C and 5% CO$_2$ for 60–70 h to allow T cells to pass two cell cycles.

14 For cell attachment and fixation continue with *Protocol 3*, step1.

4 Pre-treatments needed for hybridization

Fixed cells require further pre-treatment to obtain an efficient accessibility of the DNA probe to the nuclear target DNA. Treatment with the detergent Triton X-100 and repeated freezing in liquid nitrogen after incubation in glycerol, helps to make nuclear DNA accessible for FISH probes without strongly affecting the 3D chromatin architecture (see below, *Figure 5*, and *Plate 22*). These two steps are generally sufficient for hybridization of highly repetitive sequences, e.g. for mapping centromere regions of interphase chromosomes with alphoid DNA probes. When less repetitive or single copy DNA sequences are targeted, additional deproteinization is necessary. There are two methods of deproteinization:

(a) Incubation in HCl.

(b) Digestion with pepsin.

These pre-treatments may be combined or used separately. In many cases, a short incubation in 0.1 M HCl makes nuclear DNA sufficiently accessible for probes. Nevertheless, depending on the cell type processed, additional pepsin treatment may improve hybridization signals of, for example, cosmid probes. Pepsin incubation should be monitored under the microscope as the duration of pepsin treatment critically affects the preservation of the nuclear morphology. It is important to take into account that pepsin treatment tends to cause more

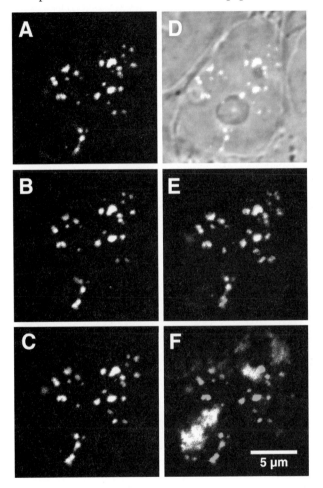

Figure 5 (See *Plate 22*) Preservation of the chromatin structure in neuroblastoma cell nucleus after fixation and performance of the 3D FISH. (A–C) Three optical sections from three different stacks of confocal serial sections through a nucleus of the same cell after replication pulse labelling using Cy3–dUTP microinjection. The nucleus demonstrates late S phase pattern. Most matching optical sections from the nucleus in living cell (A), in the same cell after fixation (B), and in the same cell after 3D FISH (C) are shown. Note the similar pattern of the replication foci, and the same shape and size on all three sections. (D) Overlay of a phase-contrast image of the nucleus in living cell and replication labelling (green) from the optical section shown in (A). (E) Overlay of optical sections from living cell (A) in green colour and optical section from the same nucleus after 3D FISH (C) in red colour. Overlay shows that replication labelled chromatin has not significantly moved or dispersed after fixation and hybridization. (F) The same optical section as in (C) with both replication labelled chromatin foci (red) and painted chromosome 5 territories (green).

129

damage in the 3D morphology of the nuclei than 0.1 M HCl treatment. If a strong background develops after DNA–DNA hybridization due to a high content of RNA in nuclei or cytoplasm, cells should be treated with RNase. RNase treatment is also necessary as a control for DNA/RNA hybridization.

Protocol 5

Post-fixation treatments

Reagents

- PBS with 0.5% Triton X-100 (Merck)
- 20% glycerol in PBS
- 0.1 M HCl in H_2O
- Liquid nitrogen[a]
- 0.002% pepsin in 0.01 M HCl (Sigma): make a 10% stock solution in H_2O, store aliquots at –20 °C

- 200 µg/ml RNase in 2 × SSC (Roche): make a 2% RNase stock solution in Tris–HCl buffer, store aliquots at –20 °C
- 1% PFA in PBS (Merck) (see *Protocol 2*)
- 50% formamide (v/v) (Merck) in 2 × SSC pH 7.0[b]

A. Obligatory steps

1 After fixation, wash slides in PBS, 3 × 5 min.

2 Incubate in PBS with 0.5% Triton X-100 for 10 min.

3 Incubate in 20% glycerol in PBS for 30–60 min (or overnight).

4 Take a slide with cells out of a Coplin jar using forceps, dip the slide into liquid nitrogen for a couple of seconds—until a characteristic 'click' sound is heard. Place the slide on a paper towel and wait until glycerol on the slide thaws. If handled one by one, slides and coverslips withstand freezing–thawing and do not break.

5 Repeat the freeze–thaw treatment three more times. Before each freezing, briefly soak slides in glycerol.

6 Wash in PBS, 3 × 5 min.

7 Incubate in 0.1 M HCl for 5–10 min. Duration of HCl treatment should be adjusted for each cell type: too short an incubation may result in a weak hybridization signal, over-incubation leads to poor nuclear morphology after heat denaturation.

8 Wash in PBS, 2 × 5 min.

B. Optional digestion with pepsin

1 Incubate in 0.002% pepsin in 0.01 M HCl at 37 °C, for 8 min.

2 Wash in PBS, 2 × 5 min.

3 Post-fix in 1% PFA/PBS, 10 min.

4 Rinse in PBS, 3 × 5 min.

Protocol 5 continued

C. Optional treatment with RNase

1 Load 200–400 µl of 200 µg/ml RNase A solution on each slide, cover with a coverslip, and incubate at 37 °C for 30 min.

2 Wash in 2 × SSC, 3 × 5 min.

D. Equilibration of fixed cells in formamide and storage

1 Equilibrate slides in 50% formamide/2 × SSC for 1–2 h at room temperature prior to hybridization. Storage in this solution for one to two weeks at 4 °C improves the hybridization efficiency. Cells may be stored this way for up to three months.

a Safety measures must be taken: keep liquid nitrogen in a Dewar container and wear safety goggles when handling it.

b Formamide is toxic, so all the steps involving the use of this reagent should be performed in a hood wearing gloves.

5 Hybridization set-up

3D FISH is more laborious, troublesome, and more sensitive to small alterations in the protocol, than FISH to metaphase chromosomes—especially with regard to the intensity of the hybridization signal and the background level. To achieve an optimal ratio of hybridization signal intensity to background, probe labelling and probe detection schemes are crucially important. Both parameters should be pre-tested in a 2D FISH experiment. We strongly recommend performing 3D FISH experiments only after a successful hybridization to metaphase chromosomes. Moreover, we recommend always running 3D FISH experiments in parallel with a 2D control performed on a slide with metaphase spreads. Below we briefly discuss those steps of the protocol that are either different between 3D and 2D FISH or especially important for a successful performance of 3D FISH.

5.1 Probe labelling

The four haptens frequently used for DNA labelling—biotin, digoxigenin, estradiol, and dinitrophenyl (DNP) (see also Chapter 2)—can be used for 3D FISH (*Figure 1A, B*; see also *Plate 19A, B*). Hapten labelled probes detected by one or two layers of antibodies provide strong hybridization signals. Non-specific binding of probe and detection antibodies can lead to background problems, in particular in cells where a lot of cytoplasm is still present around the nuclei. Directly labelled probes do not require detection and help to minimize background problems. Unfortunately, directly labelled probes often produce only a weak hybridization signal in 3D FISH experiments. Therefore directly labelled probes are presently recommended only for the detection of repetitive (e.g. alphoid) sequences (*Figure 2E*; see also *Plate 20E*).

The number of different DNA targets detected simultaneously is limited by the number of fluorochromes that can be distinguished by the available microscope equipment. With conventional microscopes equipped with narrow excitation and emission filters, up to eight fluorochromes can presently be distinguished (25). Most CLSMs, however, are not equipped to work with more than three fluorochromes simultaneously. When more than three targets are to be visualized, combinatorial probe labelling protocols can be employed: multicoloured probes are prepared by mixing of probes containing the same DNA, but different haptens (*Figure 2A*, and *Plate 20A*; see also Chapter 10). The number of targets that can be distinguished following this strategy is given by the formula ($2^n - 1$) where 'n' represents the number of dyes (26).

5.2 Probe preparation

Due to some probe dilution during the hybridization set-up (see *Protocol 6*) 3D FISH requires a two- to three-fold higher probe concentration than normally used for the respective DNA probe in 2D FISH experiments. Low probe concentration is probably the main reason for the weak signal often observed when 3D FISH is performed with commercially available probe sets that use hybridization mixtures designed for 2D FISH to metaphase spreads.

As a rule of thumb for an untested probe, we recommend using 40–60 ng/μl of probe DNA in the hybridization mixture for chromosome-specific paints or locus-specific probes (i.e. cosmids) and 1 ng/μl for centromere-specific and other highly repetitive sequences. Occasionally it may be helpful to increase probe concentrations. The concentration of unlabelled competitor DNA (e.g. Cot-1 DNA) added to the probe for suppression of non-specific hybridization depends on the presence of repetitive sequences in the probe and can be up to 50 $\times$ the concentration of the probe DNA.

With regard to combinatorially labelled probes, two points should be taken into account. First, combinatorial labelling typically leads to a lower signal intensity of each separate label, since the differently labelled sequences compete for the same target DNA. Secondly, probe fractions labelled with different haptens should be mixed in such proportions that their fluorescence intensities after detection are equal.

To avoid squashing of 3D-fixed cells by coverslips, it is necessary to use higher volumes of hybridization mixture per coverslip, than for 2D FISH. We recommend at least 5 μl for the 18 $\times$ 18 mm coverslips and 8 μl for the 22 $\times$ 22 mm coverslips.

Precipitation of probe DNA in ethanol and subsequent redissolving of the probe in the hybridization mixture (50% formamide, 10% dextran sulfate, 2 $\times$ SSC) is carried out as described in Chapter 3, *Protocol 1*.

5.3 DNA denaturation and hybridization

Cellular DNA and probe DNA can be denatured simultaneously, even in the case of probes which require high excess of Cot-1 DNA (*Protocol 6A*). Simultaneous denaturation is quick, simple, and optimal for retention of 3D morphology. We recommend avoiding separate denaturation except for cases where pre-annealing of a given probe DNA provides an extra benefit to reduce non-specific hybridization (*Protocol 6B*).

For hybridization of DNA probes to RNA targets see Chapter 6. In experiments with DNA/RNA hybridization, two control slides should always be run:

(a) The positive probe control, DNA/DNA hybridization after denaturation and RNase treatment.

(b) The negative control for the specificity of RNA hybridization, i.e. DNA/RNA hybridization after RNase digestion but no denaturation.

Treatment of slides with RNase is preferably performed before incubation of the slides in 50% formamide according to *Protocol 5C*.

Protocol 6

Probe loading, denaturation, and hybridization

Equipment and reagents

- Hot block with exact temperature control, 75 °C
- Water-baths, 37 °C and 70 °C
- 70% formamide in 2 × SSC pH 7.0[a]
- 2 × SSC pH 7.0, from 20 × SSC stock: 3 M NaCl, 0.3 M Na citrate
- Rubber cement (e.g. Fixogum)

A. Simultaneous denaturation of probe DNA and chromatin

1 Place hybridization solution containing probe on a coverslip (e.g. 18 × 18 mm).

2 Pull a slide out of the Coplin jar with 50% formamide/2 × SSC (storage solution) and quickly drain the excess of fluid off the slide.[a]

3 Cover the selected area of the slide by the coverslip with probe.

These operations should be done very quickly in order not to dry the cells. Therefore it is obligatory to process slides one by one.

4 Wipe the excess fluid around the coverslip with soft paper and seal the coverslip with rubber cement. Let the rubber cement dry completely.

5 After rubber cement is dried, place slides on a hot block with exactly regulated temperature and denature cellular and probe DNA at 75 °C for 2–4 min. It is crucial to mind the exact temperature and time. Too short a denaturation results in poor hybridization, whereas over-denaturation damages the morphology of nuclei.

6 Hybridize at least overnight or longer (two to three days) at 37 °C.[b]

Protocol 6 continued

B. Separate denaturation of probe DNA and chromatin

1   Denature probe at 75 °C for 5 min and allow it to pre-anneal at 37 °C for 30–60 min. Place the pre-annealed probe on a coverslip just before or during chromatin denaturation (step 2).

2   To denature nuclear DNA, place slides for 2–3 min in 70% formamide/2 × SSC at 70 °C in a Coplin jar.

3   Pull the slide out, drain the excess formamide, and quickly place the slide upside down (!) on the prepared coverslip with the probe. Quickly turn the slide back, seal coverslip, and carry out hybridization as described above.

[a] Formamide is toxic, so all the steps that involve the use of this reagent should be performed in a hood and gloves should be worn.

[b] We recommend incubating slides in a metallic box which is kept in a 37 °C water-bath.

6 Post-hybridization washes, detection, nuclei counterstaining, and slide mounting

6.1 Post-hybridization washes

Washing after hybridization is performed as for 2D FISH; though one should be particularly careful not to let cells dry out. Stringency of washes depends on the probe.

Protocol 7
Post-hybridization washes

Equipment and reagents

- Water-baths
- Washing solutions: 2 × SSC, 4 × SSC, 0.1 × SSC are made from 20 × SSC (see *Protocol 6*)

by dilution with ddH$_2$O and adjusting to pH 7.0

Method

1   Remove rubber cement from the coverslip using forceps, carefully lift up the coverslip, and flick it off.

2   Wash slides in 2 × SSC, 3 × 5 min at 37 °C, with shaking.

3   Wash slides in 0.1 × SSC, 3 × 5 min at 60 °C, with shaking.[a]

[a] Stringency of this wash, which is defined by temperature and molarity of SSC buffer, may vary depending on the probe used.

6.2 Detection of hybridized probes

The choice of the detection scheme depends on several factors:

(a) The number of haptens used for probe labelling.

(b) The number of antibody layers required to obtain a sufficiently strong signal.

(c) The nuclear counterstain.

Some probes show high hybridization efficiency and require only one detection layer. This is the optimal case, because the treatment is quick and background remains low. For other probes, especially for single copy sequences, two or even three layers of antibodies are necessary to obtain a sufficiently strong hybridization signal. Several proven combinations for the detection of one to three haptens and appropriate counterstains are shown in *Tables 1–3*.

Table 1 Two haptens, one layer detection

Hapten	Layer of antibodies	Fluorescence
Biotin	Avidin–Cy3 (or Cy3.5)	Red
Digoxigenin	Sheep anti-dig–FITC	Green
Counterstain (DAPI)		Blue
Biotin	Avidin–FITC	Green
Digoxigenin	Mouse anti-dig–Cy5	Infrared
Counterstain (PI)		Red

Table 2 Two–three haptens, two layers detection

Hapten	First layer	Second layer	Fluorescence
Biotin	Avidin–FITC	Goat anti-avidin–FITC	Green
Digoxigenin	Mouse anti-dig	Goat anti-mouse–Cy5	Infrared
Counterstain (PI)			Red
Example is shown in Figure 1C, D, Figure 3A–E (see also Plate 19C, D, Plate 21A–E)			
Biotin	Avidin–FITC	Goat anti-avidin–FITC	Green
Digoxigenin	Mouse anti-dig	Sheep anti-mouse–Cy3	Red
Oestradiol	Rabbit anti-oestradiol	Goat anti-rabbit–Cy5	Infrared
Counterstain (DAPI)			Blue
Example is shown in Figure 1A (see also Plate 19A)			
DNP	Goat anti-DNP	Rabbit anti-goat–FITC	Green
Biotin		Streptavidin–Cy5	Infrared
Digoxigenin		Mouse anti-dig–Cy3	Red
Counterstain (DAPI)			Blue
Example is shown in Figure 1B (see also Plate 19B)			

Table 3 Two haptens, three layers detection

Hapten	First layer	Second layer	Third layer	Fluorescence
Biotin	Avidin–Cy5	Goat anti-avidin–Bio	Avidin–Cy5	Infrared
Digoxigenin		Rabbit anti-dig	Goat anti-rabbit–FITC	Green
Counterstain (PI)				Red

Example is shown in Figure 2B (see also Plate 20B)

Hapten	First layer	Second layer	Third layer	Fluorescence
Biotin	Avidin–Cy5	Goat anti-avidin–Bio	Avidin–Cy5	Infrared
Digoxigenin		Rabbit anti-dig	Goat anti-rabbit–Cy3	Red
Counterstain (YOYO)				Green

Example is shown in Figure 1G (see also Plate 19G)

Hapten	First layer	Second layer	Third layer	Fluorescence
Biotin	Avidin–FITC	Goat anti-avidin–Bio	Avidin–FITC	Green
Digoxigenin		Mouse anti-dig	Sheep anti-mouse–Cy3	Red
Counterstain (TO-PRO)				Infrared

Protocol 8

Detection of haptens in hybridized probes

Equipment and reagents

- Light-proof humid chambers for incubation of slides with antibodies
- Coverslips larger than the hybridized area, e.g. 24 × 32 mm or 24 × 50 mm
- SSCT: 4 × SSC with 0.2% Tween 20 (Merck)
- Blocking solution: 4% bovine serum albumin (BSA) in SSCT; store the stock solution of 20% BSA in 4 × SSC with 0.04% Na azide at 4 °C
- Antibodies or avidin[a,b]

Method

1   Incubate slides in blocking solution for 10 min, then apply antibodies.

2   Antibodies and avidin are diluted in blocking solution. Allow approx. 100 μl of antibody dilution per slide. If needed, combine antibodies for each layer in an Eppendorf tube, mix well, and centrifuge at 2100 g for 5 min.

3   Apply diluted antibodies (or mixture of antibodies) to a slide, cover with a coverslip, and incubate in a dark humid chamber for 30 min at 37 °C or at room temperature.

4   Wash slides in SSCT, 3 × 3 min.

5   Steps 3 and 4 are repeated for each successive layer of antibodies.

[a] If several haptens have to be detected simultaneously a careful selection of the antibodies has to be done in order to avoid cross-reactions. Since antiserum from one species can cross-react with immunoglobulins of other species we recommend using specifically pre-adsorbed antisera if available.

Protocol 8 continued

[b] Below we list the antibodies routinely used in our laboratory and mentioned in *Tables 1–3*, with manufacturers and appropriate working dilutions that may vary for different conditions. Many other antibodies, not listed here, can also be used.

Avidin–FITC (Vector), 1:200

Avidin–Cy3 (Jackson ImmunoResearch Laboratories), 1:500

Avidin–Cy3.5, 1:500

Avidin–Cy5 (Jackson ImmunoResearch Laboratories), 1:200

Streptavidin–Cy5 (Rockland), 1:100

Goat anti-avidin–biotin (Vector), 1:200

Goat anti-avidin–FITC (Vector), 1:200

Sheep anti-dig–FITC (Roche Molecular Biochemicals), 1:100

Mouse anti-dig–Cy3 (Jackson ImmunoResearch Laboratories), 1:100

Mouse anti-dig–Cy5 (Jackson ImmunoResearch Laboratories), 1:100

Mouse anti-dig (Sigma), 1:500

Rabbit anti-dig (Sigma), 1:500

Rabbit anti-estradiol (Roche Molecular Biochemicals), 1:200

Goat anti-DNP (Sigma), 1:200

Sheep anti-mouse–Cy3 (Jackson ImmunoResearch Laboratories), 1:500

Goat anti-mouse–Cy5 (Jackson ImmunoResearch Laboratories), 1:100

Goat anti-rabbit–FITC (BioSource), 1:200

Goat anti-rabbit–Cy3 (Amersham Pharmacia Biotech), 1:500

Goat anti-rabbit–Cy5 (Amersham Pharmacia Biotech), 1:200

Rabbit anti-goat–FITC (Sigma), 1:200

6.3 Counterstaining of nuclei and mounting cells in antifade medium

Only well preserved nuclei should be analysed after 3D FISH. As a certain proportion of nuclei is always damaged, an appropriate counterstaining is required to select sufficiently preserved nuclei.

The choice of the counterstaining dye is determined by the type of microscopy planned, by the microscope equipment, and by the fluorochromes used for detection. DAPI binds to DNA, not to RNA, and is often used for chromatin staining (*Figure 3A*; see also *Plate 21A*). However, its preferential staining of AT-rich sequences should be taken into account. Propidium iodide (PI), YOYO-1, SYTO 16, TOTO-3, and TO-PRO-3 stain both DNA and RNA, and necessitate RNase treatment of cells when it is important to distinguish reliably between DNA and RNA. The standard FISH procedure usually removes most RNA, so that PI, YOYO-1, SYTO 16, and TO-PRO-3 cause only a weak cytoplasmic staining. Nucleolar RNA is removed only partially by pre-treatment with 0.1 M HCl. Therefore cells not treated with RNase usually still show bright nucleolar staining with the above mentioned dyes.

The observation of DAPI staining by a confocal laser scanning microscope (CLSM), requires excitation with UV light. UV laser equipment is expensive and available only in few laboratories, therefore, in the absence of a UV laser, PI (*Figure 2E*; see also *Plate 20E*), TO-PRO-3, TOTO-3, SYTO 16, or YOYO-1 (*Figure 1G*; see also *Plate 19G*) are the counterstaining dyes of choice for laser scanning microscopy.

Protocol 9

Counterstaining of nuclei and mounting in antifade medium

Equipment and reagents

- Microscope slides
- DAPI (4′,6-diamidino-2-phenylindole) (Sigma): 0.02 μg/ml in 2 × SSC
- Propidium iodide (PI) (Sigma): 1–10 μg/ml in 2 × SSC
- YOYO-1 (Molecular Probes): 2.5 nM in 2 × SSC (or PBS); prepare a fresh working solution each time from a stock solution (2.5 μM in DMSO), store at 4°C
- SYTO 16 (Molecular Probes): 10 μM in 2 × SSC; prepare from stock solution (1 mM in DMSO), store at 4°C
- TO-PRO-3 (Molecular Probes): 1 μM in 2 × SSC, prepare from stock solution (1 mM in DMSO), store at 4°C[a]
- Antifade mounting medium (i.e. Vectashield, Vector Laboratories)
- Colourless nail polish or so-called base coat

Method

1 Stain in a working solution of:

    (a) DAPI for 1–3 min.

    (b) PI for 1–5 min in case of denatured nuclei, or 15–30 min if chromatin was not denatured.

    (c) TO-PRO-3 for 5 min.

    (d) YOYO-1 for 30–40 min.

    (e) SYTO 16 for 10 min.

2 Briefly wash in 2 × SSC or PBS prior to mounting.

3 After washing place antifade mounting medium on the hybridized area of coverslip and mount the coverslip itself on microscope slide. Be careful not to dry cells and not to squash preparations!

4 Remove excess of antifade around the coverslip with a soft paper and seal the coverslip with nail polish.

[a] Since this stain bleaches quickly, we recommend using a high working concentration of the dye, so that nuclei are visible by eye using a red or infrared filter of an epifluorescence microscope. An intensive staining requires less laser energy during the collection of images with CSLM and thus reduces bleaching.

7 Combined 3D FISH and replication labelling

7.1 Replication labelling

FISH can be combined with replication labelling of nuclei by incorporation of halogenated nucleotides: bromodeoxyuridine (BrdU), iododeoxyuridine (IdU), or chlorodeoxyuridine (CldU). BrdU is most often used for replication labelling (*Figure 2B*; see also *Plate 20B*), while IdU and CldU are used together in double labelling protocols which allow one to distinguish between early and mid-to-late replicating chromatin in the same nucleus (27).

The duration of incubation with halogenated nucleotides depends on the specific question addressed. Here, we describe the pulse labelling of cells in S phase (*Protocol 10A*) and double labelling of early and mid-to-late replicating chromatin (*Protocol 10B*). For pulse labelling we recommend incubating cells in BrdU for not less than 30 min. For double labelling, the duration of the pulse labelling with CldU and IdU and the chase in between has to be adjusted according to the cycling time of the respective cells.

Protocol 10

Replication labelling by the incorporation of halogenated deoxyuridines

Reagents

- 5-Bromo-2′-deoxyuridine (BrdU) (Sigma): make a 1 mM stock solution in H_2O, store aliquots at −20 °C

- 5-Iodo-2′-deoxyuridine (IdU) (Sigma): make a 1 mM stock solution in 40% DMSO, store aliquots at −20 °C

- 5-Chloro-2′-deoxyuridine (CldU) (Sigma): make a 1 mM stock solution in H_2O, store aliquots at −20 °C

A. Pulse replication labelling with BrdU

1 For adherent cells, add the required amount of BrdU to the cell culture for a final concentration of 10 μM, and incubate for 30–60 min. For cells growing in suspension make sure that they are continuously incubated in BrdU until the fixation step.[a]

2 Continue with *Protocol 2*, step 3, for fixation in the case of adherent cells or with *Protocol 3*, step 3, in the case of cells growing in suspension.[b]

B. Double replication labelling by IdU and CldU

1 Add the required amount of IdU to the cell culture medium to a final concentration of 10 μM. Incubate for 30–45 min.

2 Remove the medium[c] and rinse cells twice in PBS.

Protocol 10 continued

3 Incubate cells in fresh medium for a time period equal to 1/2–3/4 of the duration of S phase (e.g. 4–6 h for fibroblasts and stimulated lymphocytes), then add the required amount of CldU to the cell culture to a final concentration of 10 μM, and incubate for 15–45 min.[d]

4 Continue with *Protocol 2*, step 3, for fixation of adherently growing cells or with *Protocol 3*, step 3, of cells growing in suspension.[b]

[a] Start BrdU incubation in conventional culture medium and continue during attachment of the cells to slides. For the latter step also add the required amount of BrdU to the 50% FCS/RPMI (see *Protocol 3*, step 5). Total duration of incubation in BrdU for pulse labelling may be up to 1 h.

[b] For cells in suspension make sure that you have ready-to-use poly-lysine-coated slides as described in *Protocol 3*, steps 1 and 2.

[c] For adherently growing cells discard the medium from the culture flask; for cells growing in suspension, transfer the cell suspension into a plastic tube and centrifuge at 170 g for 10 min, then discard the supernatant.

[d] For non-adherently growing cells, combine CldU treatment with cell attachment. Total duration of incubation in CldU should not exceed 45 min.

7.2 Detection of incorporated halogenated deoxyuridines after FISH

Both detection of replication labelling and FISH require denaturation of DNA. When combining replication labelling and 3D FISH, the treatment with HCl (*Protocol 5*) and heat denaturation necessary for hybridization (*Protocol 6*) are also sufficient for the subsequent detection of halogenated nucleotides.

Various protocols for combined detection of FISH signals and replication labelling have been published (29, 30). We obtained better results by first performing the complete FISH detection using antibodies in SSC, and then BrdU or IdU/CldU detection using antibodies in PBS (*Table 4*). Sequential detection seems to be preferable as different antibodies work best in different salt solutions. If only BrdU is to be detected, all commercially available anti-BrdU antibodies can be used (*Protocol 11A*). Some of these antibodies also bind with different affinity to epitopes of IdU and CldU. Therefore, if both IdU and CldU have to be detected simultaneously a careful selection of antibodies must be made in order to minimize cross-reaction between these two antigens (*Protocol 11B*). Different methods have been described to achieve a clear discrimination between these two labels (27, 29). We suggest applying sequentially the primary and secondary antibodies for CldU and then the primary and secondary antibodies for IdU (*Table 4*).

Table 4 Recommended sequence for detection of 3D FISH and replication labelling

Protocol 11

Detection of incorporated halogenated nucleotides after 3D FISH

Reagents

- PBS
- PBT: PBS/0.05% Tween 20
- High salt buffer: 0.5 M NaCl, 40 mM Tris–HCl, 0.05% Tween 20 pH 8
- Blocking solution: 4% bovine serum albumin (BSA) in PBT; store the stock solution of 20% BSA in PBS with 0.04% Na azide at 4 °C
- Mouse anti-BrdU No. 7580 clone B44 (Roche), dilute 1:200

- Mouse anti-IdU (Becton Dickinson), dilute 1:20
- Goat anti-mouse–Alexa488 *highly cross-adsorbed* (Molecular Probes), dilute 1:400
- Rat anti-CldU MAS 250c, clone Bu/75 (Harlan Sero-Labs), dilute 1:200
- Goat anti-rat–Cy3 (Amersham Pharmacia Biotech), dilute 1:800

Method

1 After the last washing step of FISH detection (*Protocol 8*, SSCT washing), equilibrate slides in PBT for 5 min.

2 Incubate slides in blocking solution for 10 min.

141

3 Dilute antibodies in PBT in an Eppendorf tube, mix well, centrifuge at 21 000 g for 5 min, and keep on ice in the dark until use. Allow 120 µl for each slide.

A. Detection of BrdU

1 Apply anti-BrdU antibodies (approx. 100 µl per slide) and incubate in a dark humid chamber for 30 min at room temperature or 37 °C.

2 Wash slides in PBT, 3 × 3 min.

3 Repeat steps 1 and 2 for the secondary antibody layer.

B. Detection of IdU and CldU

1 Apply antibodies against CldU[a] (rat anti-CldU), incubate in dark humid chamber for 30 min.

2 Wash in PBT, 3 × 3 min.

3 Apply secondary antibodies for CldU detection (goat anti-rat–Cy3), incubate in dark humid chamber for 30 min.

4 Wash in PBT, 3 × 3 min.

5 Apply antibodies against IdU[a] (mouse anti-IdU), incubate in dark humid chamber for 30 min.

6 Wash in PBT, 2 × 3 min, incubate in high salt buffer for 6 min,[b] and equilibrate in PBT for 3 min.

7 Apply secondary antibodies for IdU detection (goat anti-mouse–Alexa488), incubate in dark humid chamber for 30 min.

8 Wash in PBT, 3 × 3 min.

9 Perform counterstaining as described in *Protocol 9*.

[a] The recommended rat anti-CldU antibodies detect both BrdU and CldU, but not IdU, and should be applied first. The recommended mouse anti-IdU antibodies, which are used for the detection of IdU, have some affinity also to CldU and therefore must be used after the detection of CldU is completed.

[b] The high salt buffer wash is applied in order to remove mouse anti-IdU antibodies bound to free CldU epitopes.

8 Combined protein immunodetection and 3D FISH

The critical steps which may interfere with a successful combination of DNA–DNA FISH and immunodetection of nuclear proteins are pre-treatment with HCl and heat denaturation because both treatments degrade proteins. Soluble proteins seem to be more problematic for detection in combination with FISH, than filamentous proteins (9). By and large, some proteins are difficult to detect in combination with FISH, whereas some are more resistant or more abundant so that a sufficient amount of antigen withstands HCl treatment and denaturation

(*Figure 2E*; see also *Plate 20E*). Some authors carry out the detection of proteins separately before FISH (31) and some even perform immunodetection of proteins separately after hybridization and detection of FISH signals (9, 21).

Applicability of different protocols may depend upon cell type, target DNA, and proteins; each particular task may require certain modifications of the protocol. Based on our experience, we recommend starting from protein detection by applying both primary and secondary antibodies before FISH. The secondary antibodies should be preferably conjugated with stable fluorochromes like Cy3 (*Figure 2C*; see also *Plate 20C*) or with biotin (*Figure 2D*; see also *Plate 20D*) which are heat resistant. Stabilization of these antibodies by post-fixation in PFA decreases FISH signal and has proven to be not necessary. Biotin may be detected together with the detection of the hybridized DNA (*Table 5*).

Table 5 Possible sequences of combined DNA hybridization and protein detection

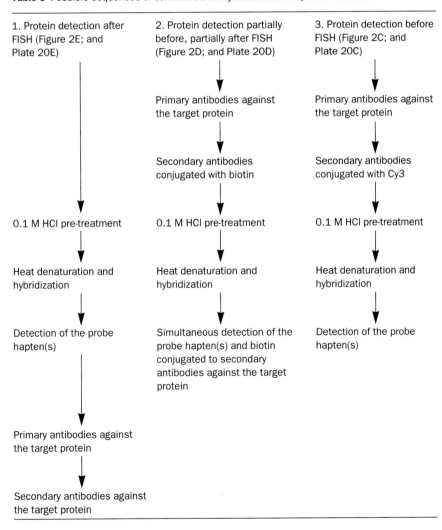

1. Protein detection after FISH (Figure 2E; and Plate 20E)	2. Protein detection partially before, partially after FISH (Figure 2D; and Plate 20D)	3. Protein detection before FISH (Figure 2C; and Plate 20C)
	Primary antibodies against the target protein	Primary antibodies against the target protein
	Secondary antibodies conjugated with biotin	Secondary antibodies conjugated with Cy3
0.1 M HCl pre-treatment	0.1 M HCl pre-treatment	0.1 M HCl pre-treatment
Heat denaturation and hybridization	Heat denaturation and hybridization	Heat denaturation and hybridization
Detection of the probe hapten(s)	Simultaneous detection of the probe hapten(s) and biotin conjugated to secondary antibodies against the target protein	Detection of the probe hapten(s)
Primary antibodies against the target protein		
Secondary antibodies against the target protein		

143

9 Preservation of the chromatin structure during 3D FISH

The preservation of the nuclear morphology is essential for the interpretation of 3D FISH experiments, which require harsh successive treatments (detergents, freeze–thaw steps, HCl, and heat denaturation) after fixation. This raises the question of the structural preservation of the nucleus after 3D FISH. So far there are only a few experimental data addressing this problem. It was shown that in PFA-fixed nuclei, centromere regions detected by a kinetochore-specific antibody did not change their positions after 3D FISH and remained co-localized with centromere-specific DNA.

Assessment of chromatin topology in the nucleus of a living cell has recently become feasible, based on the *in vivo* visualization of chromatin using replication labelling with fluorochrome labelled nucleotides (30, 32, 33). This approach yields living cells with nuclei exhibiting a number of chromosome territories built up from fluorescently labelled subchromosomal foci. The approach has allowed us, for the first time, to study the 3D topology of chromatin domains in the nucleus of one and the same cell in the living state (*Figure 5A, D*; see also *Plate 22A, D*), after fixation (*Figure 5B*; see also *Plate 22B*), and after 3D FISH (*Figure 5C, F*; see also *Plate 22C, F*). In this way we have tested the extent to which this 3D topology could be preserved using the 3D FISH protocol described in this chapter. Visual comparison of confocal optical sections through individual nuclei observed at the living, fixed, and hybridized stages showed that the chromatin domains retained their positions, shape, and size both after fixation with 4% paraformaldehyde and after accomplishment of FISH (compare *Figure 5A, B*, and *C*; and also *Plate 22A, B*, and *C*). This indicates that qualitative and quantitative observations performed after 3D FISH reflect the situation *in vivo* down to individual chromatin domains reasonably well (Solovei *et al.*, manuscript in press).

Fixation with methanol/acetic acid (3:1)—commonly used for cytogenetic studies—provides bright signals after FISH on cells grown and fixed on slides. However, such fixation fails to preserve the morphological integrity of the nuclei after heat denaturation, as noted after DNA counterstaining, and we do not recommended this fixation procedure for 3D FISH.

10 Confocal microscopy

A confocal laser scanning microscope (CLSM) is often used for the 3D analysis of nuclei subjected to 3D FISH protocols, although a conventional fluorescence microscope equipped with a CCD camera may also be used, if certain requirements can be met (see below). The basic understanding of the optical principles and of the advantages and limitations of different types of far field fluorescence microscopy is indispensable when using such equipment for the evaluation of 3D FISH experiments. Note that technical demands for a 3D data set useful for quantitative 3D image analysis, e.g. for distance measurements between closely spaced nuclear signals, are much higher than for applications that do not approach the limits of light microscopic resolution. For this reason a brief account of the problems of quantitative 3D microscopy is included in this chapter with a focus on practical hints. In contrast to a conventional epifluorescence microscope, which forms a blurred image including a wide range above and below the focus plane, the confocal microscope image contains information from only a small range. Light emerging from above or below the plane of focus is excluded by means of a pinhole. Information about the three-dimensional distribution of dye concentration (fluorescence intensity) in a specimen can be collected by taking a series of images ('light optical sections') at different axial positions, a so-called 'image stack'. In this stack, the images are separated from each other in an axial direction by the 'axial step width' (or 'z-step width'). This procedure yields a three-dimensional array of data elements ('voxels') for image analysis (see below).

The principle of the confocal microscope is inherently different from that of a wide field microscope, which forms a real optical image in the camera or in the eye. In a confocal microscope (*Figure 6*) the object is scanned sequentially with a laser focus point by point and the fluorescence signal emitted by each point is recorded by a sensor, which is in most systems a photomultiplier tube (PMT). The image is then assembled from the recorded fluorescence intensities that are stored in computer memory. The laser excites fluorochrome molecules in a conical region throughout the specimen. All molecules in this region emit fluorescence light and would in principle contribute to the recorded signal. However, emission light from molecules outside the laser focus is strongly excluded by the pinhole positioned in front of the detector. This pinhole is confocal to the laser

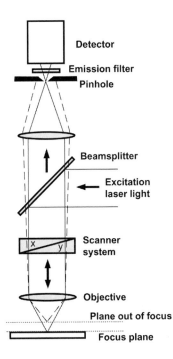

Figure 6 Scheme of the light path of a beam scanning confocal microscope. The excitation light is coupled into the light path by the beam splitter and scanned across the specimen by the scanner system. Solid lines represent the light path for one position of the scanner system. The excitation light is focused on a point in the specimen. Emission light emerging from this point passes the beam splitter, is focused on to the pinhole, and passes it. Light from planes out-of-focus (dashed lines) hits the pinhole defocused. Therefore only a small portion of this light passes the pinhole.

focus (hence the term 'confocal' microscope), i.e. light that is emitted from molecules in the laser focus is focused onto the pinhole by the optics of the microscope. Light emitted from molecules outside the laser focus does not hit the pinhole or hits it defocused, so that its contribution to the recorded signal is strongly reduced.

Scanning of the specimen can be performed in two different ways. In stage scanning the specimen is moved and the laser focus stays fixed. In beam scanning the specimen position is fixed and the laser focus is moved by means of galvanometer driven mirrors in the light path as used in our laboratory. Most modern microscopes use beam scanning as it allows for faster scanning. However it goes along with small optical aberrations, which result from moving the laser focus away from the optical axis. There are several variants of the beam scanning technique, e.g. the use of acousto optical beam deflectors for scanning (36) or the use of a rotating disc for scanning multiple beams over the specimen simultaneously (37), which will not be considered here.

The need for scanning the specimen imposes particular constraints on confocal microscopes. Take as an example a single light optical section with 512×512 pixels that is recorded in one second. This leaves for each pixel a maximum dwell time of 3.8 μsec, i.e. the time for which the laser beam illuminates a given point of the specimen. In practice, delay times in the instrument make the dwell time even shorter. Thus all photons which contribute to each pixel have to be collected within microseconds. For obtaining a reasonable fluorescence signal one has to use high excitation intensities of the order of 1 MW/m^2 or more, in a diffraction limited spot. Such highly intense, well focused monochromatic light currently can only be delivered by lasers, which makes the instrument expensive and limits, if only for funding reasons, the number of available excitation wavelengths. Even with such a high intensity of the excitation light the signal in the pixels of highest intensity rarely will consist of more than 1000 photons (38) yielding a signal-to-noise ratio (S/N ratio) of 30. In wide field microscopy every pixel of the CCD camera collects light for the whole exposure time, thus requiring much lower excitation light levels, which can be delivered by mercury arc or xenon arc lamps. Still during image capture one CCD pixel easily collects 10 000 and more photons, yielding an S/N ratio of 100 or more.

Lately there has been a growing interest in two-photon microscopy (39). This technique uses the fact that two-photon fluorochrome excitation requires high light intensities, which are only obtained in the focal region, thus achieving an inherent optical sectioning effect even without a detection pinhole and reducing photobleaching outside the plane of focus. These properties make it specifically useful for thick specimens. Whether two-photon microscopy also has advantages for thin specimens (in the range of 10 μm) is still being discussed (40).

Three-dimensionally resolved data can also be gained from a series of focal planes recorded by conventional epifluorescence microscopy throughout the specimen. As the epifluorescence microscope has no optical sectioning capability, in these images blurred signal from structures outside the focal plane adds to signal from within the focal plane leading to a low resolution in the axial direction. This blurred signal can be removed by methods known as deconvolution (41). Set-ups for deconvolution microscopy are usually less expensive than confocal microscopes, and the method has advantages for multicolour FISH where a set of fluorochromes requires a wide range of excitation wavelengths for image acquisition or in cases where short recording times are necessary, e.g. in living cell observations. However, deconvolution can give rise to particular artefacts, which have to be recognized by the user (42). Deconvolution can also be done on confocal image stacks to further increase the resolution two-fold and more (see below).

10.1 Selection of the filter configuration

The combination of excitation laser line, beam splitter, and emission filter determines how efficiently a specific fluorochrome is excited and detected. The degree to which a fluorochrome is excited by light of a specific wavelength is shown by its excitation spectrum (*Figure 7A*). Likewise, the emission spectrum of a fluorochrome shows the proportion of its emitted light that has a specific wavelength. The same reasoning applies to the transmission spectra of beam splitters and emission filters: they show what percentage of light of a specific wavelength can pass the filter. Usually fluorochrome spectra are presented

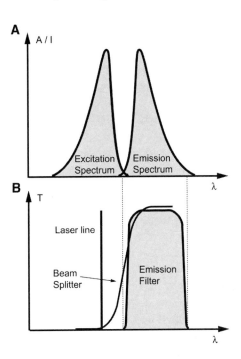

Figure 7 (A) Schematic excitation and emission spectrum of a fluorochrome. Relative absorption/emission light intensity (A/I) are plotted versus wavelength (λ). (B) Spectrum of excitation laser line, dichroic beam splitter, and emission filter matching the fluorochrome spectra. Transmissions of beam splitter and emission filter (T) are plotted versus wavelength (λ). The line represents the wavelength of the excitation laser.

normalized, i.e. their maximum is arbitrarily set to one and all other values are scaled accordingly. To make the fluorochrome spectra 'absolute', i.e. comparable to the spectra of other fluorochromes, one has to know the extinction coefficient and the quantum yield of the specific fluorochromes. Transmission values of beam splitters and filters for each wavelength are given in numbers between 0 and 1, a value of 0.6 means that 60% of the light of this wavelength is transmitted.

Selection of laser lines, beam splitters, and emission filters for a particular fluorochrome should be performed with two concerns in mind:

(a) To gain as much of the emission light of the respective fluorochrome, as possible, in the photomultiplier.

(b) To exclude as much fluorescence light from all other fluorochromes present in the specimen as possible to avoid cross-talk between the colour channels.

To obtain maximum excitation of a fluorochrome the excitation laser wavelength has to be set at the peak of the excitation spectrum. Usually there will be just one, or at most two, laser lines that are within the excitation spectrum, and usually none of these lines reflects the maximum of the spectrum. In most cases, however, one can simply counter a lower excitation efficiency by setting the laser power higher and in this way can obtain the same amount of emission photons without increasing photobleaching.

The proper choice of the emission filter has a decisive impact on the attainable signal. All fluorochromes currently used in FISH experiments undergo bleaching during image recording procedures. Thus, every emitted photon that is blocked by the emission filter reduces the maximum number of photons that contribute to the attainable signal. *Figure 7B* shows an emission filter spectrum that perfectly matches the emission spectrum of a fluorochrome. If multiple fluorochromes are recorded, narrower emission filters might have to be used to avoid cross-talk between the colour channels. For more information on fluorescence filters see ref. 43.

The function of the beam splitter is to separate emission light from excitation light. It functions as a mirror that reflects the excitation light and transmits the emission light (*Figure 6*). Presently available beam splitters do not fulfil this requirement perfectly. Even at maximal transmission a beam splitter reflects some 2% of the light. Likewise even at maximal reflection, a considerable quantity of light is transmitted. In particular back scattered excitation light passes through the beam splitter and has to be rejected by the emission filter.

As the beam splitter reflects the excitation light, even small changes in its angle towards the light path can introduce shifts in the image. Thus, to minimize such shifts when recording images from different fluorochromes sequentially, if possible, the beam splitter should not be changed. (Note: When images from different fluorochromes are recorded simultaneously, the beam splitter cannot be changed.) This requires a beam splitter that reflects all involved laser wavelengths and transmits as much of the emission light as possible. Fortunately, even a beam splitter that is not specified for a required excitation wavelength will often work for this wavelength, as even at maximal transmission at this wavelength it still reflects a small percentage of the light (an advantage of the otherwise deplorable imperfection of beam splitters).

10.2 Conditions of image acquisition

Mismatches in refractive index (RI) and coverslip thickness can introduce spherical aberrations, which blur the image considerably. Oil immersion objectives are usually corrected for a coverslip thickness of 170 μm (No. 1.5 coverslips). We have already recommended (see Section 2.1, *Protocol 1*) growing cells directly on coverslips specially selected for a narrow range of this thickness (*Figure 4*) (44). A situation in which the RI of the mounting medium matches that of the immersion oil (~1.51) is ideal. We use routinely a glycerol-based mounting medium (RI = 1.44). Water-based mounting medium is best used in combination with a water immersion objective.

The diameter of the confocal pinhole determines the resolution of the confocal microscope in the axial direction. It also influences the signal intensity, as a small pinhole excludes more light than a large one. An optimal trade-off between the axial resolution and signal intensity is to use a pinhole size of one Airy Unit (45) or slightly smaller. If more signal or larger optical sections are needed, the pinhole should be chosen larger (46). To avoid so-called 'sampling artefacts' the voxel size, i.e. the pixel size of a confocal image plane and the step width, should be smaller than the optical resolution by a factor of at least 2.3 (47). For an objective with an NA of 1.4 this translates into a maximal pixel size of 80 nm and a maximal step width of 250 nm.

When choosing the image size, it makes sense to leave several pixels of empty space between the border of the recorded specimen image and the border of the image frame. Usually one will do this intuitively as one wants to be sure not to cut off any information by choosing too small an image frame. In contrast to this, it can easily happen that one cuts off the upper and lower edges, i.e. in the axial direction, as these seem to be image planes containing almost no information. However, the information in these planes is as valid as the information in the lateral edges of the specimen. In some cases one might get the information one needs by just taking a few sections or even a single section through the centre of the object. In all cases, however, where one wants to perform a 3D reconstruction or a deconvolution of a whole, serially sectioned object, one needs the intensity information from this whole object. Therefore we recommend as a take home message: record too many sections above and below the object rather than too few.

A low S/N ratio leads to a high error of intensity measurements and limits the effective resolution, because in cases of high noise there is no way of telling whether intensity changes between neighbouring voxels reflect changes in the specimen itself, or are simply due to this noise (46). The principal cause of noise in confocal microscopes is the statistical nature of the fluorescence process. The noise varies with the square root of the signal and the only way of increasing the

S/N ratio is to increase the amount of collected photons from the specimen. For a given microscope set-up there are three ways of doing this: increasing the excitation power, scanning at a lower speed, and recording several frames of one image and averaging these frames.

The photomultiplier tube (PMT) voltage and the amplifier offset and gain determine the grey level in the image that will result from a specific signal intensity. For 8-bit digitizers the grey level is between 0 and 255. Intensities above or below the range are clipped to 0 or 255 ('in saturation'). For quantitative measurements saturation needs to be avoided. This can be achieved by displaying the images with a colour map that assigns special colours to the values 0 and 255 (so-called glow over/under or range indicator colour maps) and adjusting the PMT voltage and amplifier settings so that pixels with these special colours do not appear.

10.3 Calibration of the instrument

When recording preparations with multiple fluorochromes one has to keep in mind that the position at which a labelled structure appears in the image stack can vary depending on the fluorochrome used for labelling. This circumstance is called 'chromatic shift'. To the observer it is most apparent when a simple structure like a polystyrene bead, that is labelled with multiple fluorochromes, is imaged. The preparation of such slides for PSF (point spread function) and chromatic shift measurements is described in *Protocol 12*. The images of the bead in the different colour channels will appear shifted with respect to each other in the image stack (*Figure 3F, G*; see also *Plate 21F, G*). The chromatic shift is usually biggest in the *z*-direction. For UV excited fluorochromes it can be up to several micrometers, and even in a well aligned microscope there are often shifts of several hundred nanometers. The chromatic shift is not the same for all objectives of a given type and needs to be measured experimentally for optimal image correction. It also varies to some degree with the position in the field of view ('pan region'), which is under observation and the distance of a recorded object from the surface of the coverslip (48). To obtain the most precise measurements of the distance between two point-like signals visualized with different fluorochromes (e.g. cosmid signals of individual genes or intensity gravity centres of larger objects such as two chromosome band domains), one should measure the chromatic shift for one specific field of view for one objective lens and use the same settings for all subsequent measurements (see below, high precision distance measurements). For best reliability a shift measurement should be done for each day that biological preparations are being scanned (*Protocol 13*).

10.4 Visualization

The simplest way of displaying a confocal image stack is to make a gallery, i.e. to display all *z*-sections side by side in one image. Usually when a gallery is made, the individual images are downscaled, so that the gallery does not get too large. This means, however, that resolution is lost.

Protocol 12

Preparing slides with fluorescent beads

Equipment and reagents

- Coverslips with even thickness of 0.17 mm
- Microscope slides
- Appropriate suspension of fluorescent beads (Tetra Speck 500, Molecular Probes: T-7281)
- Mounting medium as used for biological specimens
- Nail polish for sealing coverslip to slide

Method

1   Clean dust from coverslip.

2   Shake a tube containing the beads to make a homogeneous suspension.

3   Drop 5 µl of bead suspension onto the centre of the coverslip.

4   Spread the suspension over the coverslip with the pipette tip as far as the surface tension permits.

5   Cover the coverslips containing the beads to protect them from light and dust and let them air dry (approx. 1 h).

6   Mount in the appropriate mounting medium on glass slides.

7   Seal with nail polish.

Protocol 13

Measurement of the chromatic shift

Equipment

- Slide with mounted multispectral beads (Tetra Speck 500, Molecular Probes: T-7281)

Method

1   Find a region on the slide where the beads are dense enough that several beads fit into one image. The beads should still be separated from each other by at least two bead diameters.

2   Take an image stack of the colour channels of interest and the normally used voxel size. Use a high number of averages to get a high S/N ratio. Collect about 15 beads altogether.

3   Use an evaluation software that calculates the centres of mass of the beads for each colour channel. Alternatively, if such software is not available, determine the positions of the bead centres in some visualization software by eye. Determine the position of the bead images in each colour channel separately. The result will not be as accurate but might be sufficient for your application.

4   Calculate the distance between the colour channels in *x*-, *y*-, and *z*-direction for each bead separately and take the average over all beads. Calculate the standard deviation to judge the accuracy of the measurement.

Navigation through a confocal image stack can be facilitated by a 3D-slicer, which simultaneously displays sections in three different directions (*xy*, *xz*, *yz*) through one point. By clicking on one point in a section the positions of the other two sections are changed, accordingly.

Another way to generate a quick overview of a confocal image stack is to make a projection, i.e. to map data from the whole image stack to a single image plane. Each pixel in the projection is calculated from the voxels along an imagined line through this pixel and perpendicular to the projection. In a maximum intensity projection (MIP, *Figure 3D*; see also *Plate 21D*) only the voxel with maximum intensity along the line is taken, leading to a crisp image but leaving out all the other information in the image. In an 'extended focus' or 'sum' projection, the projected pixel is the average of all voxels along the line.

Similar to doing projections but more contrived is volume rendering. Volume rendering techniques simulate a light source that illuminates the image stack. The virtual light is reflected, diffracted, absorbed, or produces fluorescence to a degree dependent on the grey levels of the voxels it hits. The exact way this works depends on the particular algorithm used and the parameters chosen for this algorithm. Finally, the image for display is calculated as an imagined observer would see this simulated scene.

Surface rendering algorithms visualize the specimen by means of a surface ('isosurface') or several surfaces, which envelop all voxels with intensities above a user defined threshold and exclude voxels with intensities below this threshold. This procedure rejects all information about structures inside and outside the surfaces. It creates the impression of a rigid structure with clear borders, which might be justified or misleading depending on the studied specimen (*Figure 3E*; see also *Plate 21E*). The result of surface rendering depends on the selected threshold. Surface rendering is less computationly intensive than volume rendering, and the direction of the view can be changed in real time on standard 3D graphics hardware, if the software supports it.

10.5 Quantitative measurements and deconvolution

Distance measurements are basically simple: Locate the first point, locate the second point, and calculate the distance between the two points by means of the

Pythagorean theorem. The actual procedure will depend on the kind of software that is available. For manual localization of the two points it is convenient to have a program that facilitates navigating through the image stack and displays the coordinates of a selected point and immediately calculates the distance between two selected points. Since two structures, which are point-like in the recorded specimen, appear as elongated ellipsoids in the image stack, distances should be determined between their image gravity centres. This requires an algorithm that can discern objects. The simplest way in which such algorithms work is by setting a global threshold value. Every coherent region of voxels with a grey level above the threshold is then taken as a separate object, of which the gravity centre is calculated. Problems arise if the chosen threshold is too low so that two separate objects are falsely counted as one object or if the threshold is too high so that some objects are completely lost. There are cases where it is not possible to choose any threshold that separates all objects from each other without losing low intensity objects. Algorithms exist, that use more contrived ways of discerning objects like application of local thresholds or detection of edges.

If distances between closely adjacent objects of a different spectral signature (i.e. labelled with different fluorochromes) should be measured, precise determination of the chromatic shift is crucial (see *Protocol 13*). The image stack recorded in one colour can then be moved by this value to match properly the image stack recorded in the other colour. Alternatively, correction of the chromatic shift can be performed after calculation of the 3D position of the intensity gravity centres by subtracting the shift from the object coordinates. The second way of correction is easier and has the advantage that it can provide accurate measurements even for distances smaller than one pixel, a procedure called spectral precision distance microscopy (SPDM) (49). In case that one has to correct for the shift directly in the image data, one will usually do this by shifting the image stack of one colour channel by the appropriate amount of voxels. A higher accuracy of the measurement can in principle be achieved by shifting the data by distances smaller than one voxel using Fourier transformation methods.

The simplest and most common way to measure the volume of objects is to apply a threshold value and measure the volume as the number of all coherent voxels with intensities above the threshold. The result of this measurement depends on the applied threshold. The measured volume might either be too high, if background voxels are classified to belong to the object, or too low, if low intensity regions of the object are cut off. If the real object is uniformly labelled with fluorochromes, this dependency of the measured volume on the threshold is caused by the blur, which is introduced by the microscopic image formation. The blur can be removed to some degree by deconvolution, which reduces the threshold dependency of the measured volume. When the real object is not uniformly labelled, but has regions of stronger labelling and regions of weaker labelling, the threshold dependency of the measured volume remains even after deconvolution. In this case, which is the more likely one for biological specimens, e.g. chromosome territories, the borders of the object tend to blend into the background and the choice of threshold becomes somewhat arbitrary. Thus, absolute volume and surface measurements are questionable and should, if at all, be performed over a range of reasonable thresholds taking the mean from this range (10, 16). More meaningful are relative comparisons of volumes, e.g. from different chromosome territories, if these territories can be painted rather homogeneously with the same fluorochrome and exhibit similar amounts of emission photons per unit volume. Since these requirements cannot be ideally met, quantitative measurements of chromosome territory volumes and surfaces have to be interpreted with caution.

The resolution of confocal fluorescence microscopy is limited by the wavelength of visual light to about 200 nm laterally and 600 nm axially. This limitation shows up in the width of the point spread function (PSF), the microscopic image of a single point-like object. In fact any object can be regarded as the composition of all its points. In this view the microscope image is formed by all PSFs of the object points. To calculate the image, the PSF would have to be moved to each point in the object, scaled with the intensity (the fluorochrome concentration in the object) of this point, and then the contributions of all PSFs to each point would have to be summed up. Such a mathematical operation is called convolution. Algorithms to reverse this operation are called deconvolution algorithms. Deconvolution algorithms increase the resolution of a microscopic image by partially removing the blur caused by the PSF (*Figure 8*). A variety of deconvolution methods exist. For an overview and a discussion of capabilities and possible artefacts see ref. 42. For deconvolution of confocal microscope data with its often very limited signal-to-noise ratio, iterative restoration algorithms like the maximum likelihood method should be used, which simultaneously increase the resolution and decrease the noise.

Image processing software is offered by many companies. Fewer products are able to handle 3D data, and even fewer have all the features one would like for the analysis of 3D FISH experiments. The list of software given in Appendix names the most common freeware and shareware tools that are used for format conversion, image display, and at least partial 3D image analysis.

References

1. Lichter, P., Cremer, T., Borden, J., Manuelidis, L., and Ward, D. C. (1988). *Hum. Genet.*, **80**, 224.
2. Manuelidis, L. (1984). *Proc. Natl. Acad. Sci. USA*, **81**, 3123.
3. Manuelidis, L. (1985). *Hum. Genet.*, **71**, 288.
4. Manuelidis, L. (1990). *Science*, **250**, 1533.
5. Manuelidis, L. and Borden, J. (1988). *Chromosoma*, **96**, 397.
6. Pinkel, D., Landegent, J., Collins, C., Fuscoe, J., Segraves, R., Lucas, J., *et al.* (1988). *Proc. Natl. Acad. Sci. USA*, **85**, 9138.
7. Rappold, G. A., Cremer, T., Hager, H. D., Davies, K. E., Muller, C. R., and Yang, T. (1984). *Hum. Genet.*, **67**, 317.
8. Schardin, M., Cremer, T., Hager, H. D., and Lang, M. (1985). *Hum. Genet.*, **71**, 281.
9. Bridger, J. M., Herrmann, H., Münkel, C., and Lichter, P. (1998). *J. Cell Sci.*, **111**, 1241.

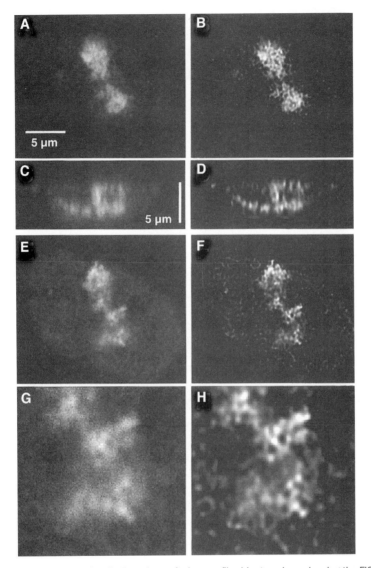

Figure 8 Central optical sections of a human fibroblast nucleus showing the FISH signal of the chromosome 11 territories detected with Cy5, recorded on a Leica TCS SP. (A, C, E, G) Sections of the original recorded data. (B, D, F, H) Corresponding sections of the data deconvolved with the MLE Algorithm of Huygens2 (Bitplane, Zürich CH) and a measured PSF. (A, B) and (E, F) Central *xy* sections. (C, D) A *yz* section. (G, H) One of the chromosome territories of (E, F) magnified.

10. Dietzel, S., Schiebel, K., Little, G., Edelmann, P., Rappold, G. A., Eils, R., *et al.* (1999). *Exp. Cell Res.*, **252**, 363.
11. Eils, R., Bertin, E., Saracoglu, K., Rinke, B., Schrock, E., Parazza, F., *et al.* (1995). *J. Microsc.*, **177**, 150.
12. Kurz, A., Lampel, S., Nickolenko, J. E., Bradl, J., Benner, A., Zirbel, R. M., *et al.* (1996). *J. Cell Biol.*, **135**, 1195.

13. Zirbel, R. M., Mathieu, U. R., Kurz, A., Cremer, T., and Lichter, P. (1993). *Chromosome Res.*, **1**, 93.

14. Bridger, J. M. and Lichter P. (1999). In *Chromosome structural analysis: a practical approach* (ed. W. A. Bickmore), p. 209, Oxford University Press.

15. Cremer, T., Kurz, A., Zirbel, R., Dietzel, S., Rinke, B., Schröck, E., *et al.* (1993). *Symp. Quant. Biol.*, **58**, 777.

16. Dietzel, S., Jauch, A., Kienle, D., Qu, G., Holtgreve-Grez, H., Eils, R., *et al.* (1998). *Chromosome Res.*, **6**, 25.

17. Solovei, I., Kienle, D., Little, G., Eils, R., Savelyeva, L., Schwab, M., *et al.* (2000). *Genes Chromosomes Cancer*, **29**, 297.

18. Lesko, S. A., Callahan, D. E., LaVilla, M. E., Wang, Z. P., and Ts'o, P. O. (1995). *Exp. Cell Res.*, **219**, 499.

19. Nagele, R. G., Freeman, T., McMorrow, L., Thomson, Z., Kitson-Wind, K., and Lee, H. (1999). *J. Cell Sci.*, **112**, 525.

20. Popp, S., Scholl, H. P., Loos, P., Jauch, A., Stelzer, E., Cremer, C., *et al.* (1990). *Exp. Cell Res.*, **189**, 1.

21. Croft, J. A., Bridger, J. M., Boyle, S., Perry, P., Teague, P., and Bickmore, W. A. (1999). *J. Cell Biol.*, **145**, 1119.

22. Dietzel, S., Eils, R., Satzler, K., Bornfleth, H., Jauch, A., Cremer, C., *et al.* (1998). *Exp. Cell Res.*, **240**, 187.

23. Tashiro, S., Kotomura, N., Shinohara, A., Tanaka, K., Ueda, K., and Kamada, N. (1996). *Oncogene*, **12**, 2165.

24. Funderud, S., Nustad, K., Lea, T., Vartdal, F., Gaudernack, G., Stenstad, P., *et al.* (1987). In *Lymphocytes: a practical approach* (ed. G. Klaus). IRL Press, Oxford.

25. Azofeifa, J., Fauth, C., Kraus, J., Maierhofer, C., Langer, S., Bolzer, A., *et al.* (2000). *Am. J. Hum. Genet.*, **66**, 1684.

26. Nederlof, P. M., van der Flier, S., Wiegant, J., Raap, A. K., Tanke, H. J., Ploem, J. S., *et al.* (1990). *Cytometry*, **11**, 126.

27. Aten, J. A., Stap, J., Manders, E. M., and Bakker, P. J. (1993). *Eur. J. Histochem.*, **37** Suppl. 4, 65.

28. Aten, J. A., Stap, J., Hoebe, R., and Bakker, P. J. (1994). *Methods Cell Biol.*, **41**, 317.

29. Visser, A. E., Eils, R., Jauch, A., Little, G., Bakker, P. J., Cremer, T., *et al.* (1998). *Exp. Cell Res.*, **243**, 398.

30. Zink, D., Cremer, T., Saffrich, R., Fischer, R., Trendelenburg, M. F., Ansorge, W., *et al.* (1998). *Hum. Genet.*, **102**, 241.

31. Ersfeld, K. and Stone, E. M. (2000). In *Protein localization by fluorescence microscopy: a practical approach* (ed. V. J. Allan), p. 51. Oxford University Press, Manchester.

32. Manders, E. M., Kimura, H., and Cook, P. R. (1999). *J. Cell Biol.*, **144**, 813.

33. Zink, D. and Cremer, T. (1998). *Curr. Biol.*, **8**, R321.

34. Robbins, W. A., Segraves, R., Pinkel, D., and Wyrobek, A. J. (1993). *Am. J. Hum. Genet.*, **52**, 799.

35. Vourc'h, C., Taruscio, D., Boyle, A. L., and Ward, D. C. (1993). *Exp. Cell Res.*, **205**, 142.

36. Tsien, R. Y. and Bacskai, B. J. (1995). In *Handbook of biological confocal microscopy* (ed. J. B. Pawley), p. 459. Plenum Press, New York.

37. Kino, G. S. (1995). In *Handbook of biological confocal microscopy* (ed. J. B. Pawley), p. 155. Plenum Press, New York.

38. Pawley, J. B. (1995). In *Handbook of biological confocal microscopy* (ed. J. B. Pawley), p. 19. Plenum Press, New York.

39. Denk, W., Piston, D. W., and Webb, W. W. (1995). In *Handbook of biological confocal microscopy* (ed. J. B. Pawley), p. 445. Plenum Press, New York.

40. Patterson, G. H. and Piston, D. W. (2000). *Biophys. J.*, **78**, 2159.

41. Shaw, P. J. (1995). In *Handbook of biological confocal microscopy* (ed. J. B. Pawley), p. 373. Plenum Press, New York.

42. McNally, J. G., Karpova, T., Cooper, J., and Conchello, J. A. (1999). *Methods*, **19**, 337.

43. Reichman, J. (2000). *Handbook of optical filters for fluorescence microscopy*. Chroma Technology Corp. An Employee-Owned Company, Vermont.

44. Sheppard, C. J. R. and Török, P. (1997). *J. Microsc.*, **185**, 366.

45. Sandison, D. R., Williams, R. M., Wells, K. S., Strickler, J., and Webb, W. W. (1995). In *Handbook of biological confocal microscopy* (ed. J. B. Pawley), p. 39. Plenum Press, New York.

46. Stelzer, E. H. K. (1998). *J. Microsc.*, **189**, 15.

47. Webb, R. H. and Dorey, C. K. (1995). In *Handbook of biological confocal microscopy* (ed. J. B. Pawley), p. 55. Plenum Press, New York.

48. Edelmann, P. and Cremer, C. (2000). *Proc. SPIE*, **3921**, 313.

49. Edelmann, P., Esa, A., Hausmann, M., and Cremer, C. (1999). *Optik*, **4**, 194.

Chapter 8

Comparative genomic hybridization on metaphase chromosomes and DNA chips

Stefan Joos, Carsten Schwänen, and Peter Lichter

Abteilung 'Organisation komplexer Genome', Deutsches Krebsforschungszentrum Heidelberg, Im Neuenheimer Feld 280, D-69120 Heidelberg, Germany.

1 Introduction

Comparative genomic hybridization (CGH) allows a genome-wide screening of chromosomal imbalances without prior knowledge of genomic regions of interest and independent of the availability of metaphase cells from the specimens to be investigated. Since its first description (1) CGH became a powerful alternative to chromosome banding and fluorescence *in situ* hybridization (FISH) analyses (2–4) and has contributed significantly to the current knowledge of genomic alterations in a large number of tumour entities (for an overview see refs 5 and 6).

As is indicated in *Figure 1* (see also *Plate 23*), for CGH analysis whole genomes of tumour cells and control cells are used as probes for FISH against chromosomes of normal metaphase cells. By comparison of signal intensities along hybridized chromosomes, relative copy number changes of chromosomal regions within the tumour genome can be identified. Dedicated software applications for digital image analysis are commercially available allowing an evaluation of CGH experiments in an objective manner.

A major advantage of CGH as compared to other cytogenetic methods (chromosome banding, FISH) is the fact that instead of metaphase chromosomes or interphase nuclei only genomic DNA is required from the cells to be studied. Best CGH results are obtained if DNA is prepared from fresh or frozen tissue samples. However, it is also possible to analyse paraffin-embedded tissue samples and small populations or single tumour cells. Small samples require physical enrichment of tumour cells, e.g. by cell sorting or micromanipulation, in combination with a representative amplification of genomic DNA by universal PCR.

Although the protocol for CGH appears very similar to FISH protocols described for the analysis of specific chromosomal loci (see *Chapter 3* of this book) considerable effort is needed to establish CGH even in laboratories experienced

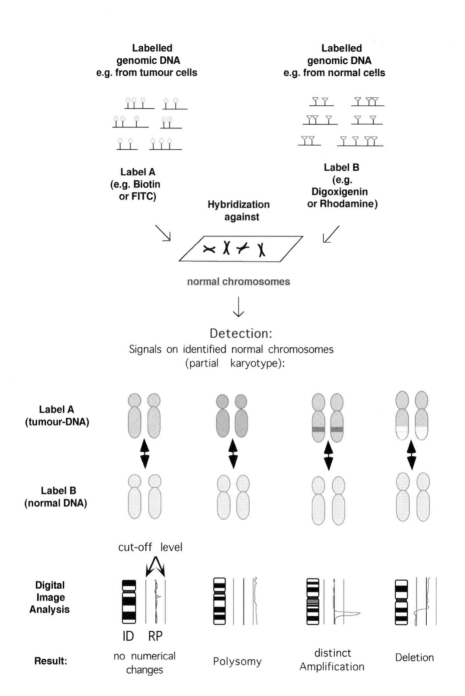

Labelled genomic DNA e.g. from tumour cells

Labelled genomic DNA e.g. from normal cells

Label A (e.g. Biotin or FITC)

Label B (e.g. Digoxigenin or Rhodamine)

Hybridization against

normal chromosomes

Detection:
Signals on identified normal chromosomes (partial karyotype):

Label A (tumour-DNA)

Label B (normal DNA)

cut-off level

Digital Image Analysis

ID   RP

Result:

| no numerical changes | Polysomy | distinct Amplification | Deletion |

in molecular cytogenetics. The success seems to be dependent on a summation of numerous steps in the protocol that have to be specially taken care of, rather than on one or two specific parameters only. The CGH protocol is described and discussed under the following headings:

- Preparation of metaphase chromosomes
- Preparation of genomic DNA from frozen tissues, paraffin-embedded material, and small tumour cell populations
- Microdissection of single cells
- Universal polymerase chain reaction (PCR) methods for the amplification of genomic DNA from single cells
- Probe labelling
- *In situ* hybridization and detection
- Image acquisition and evaluation
- Troubleshooting of CGH experiments
- Matrix CGH (CGH to microarrayed DNA)

2 Preparation of metaphase chromosomes

One of the most critical parameters for successful CGH is the quality of metaphase spreads, which should be largely devoid of residual cell debris. Furthermore, optimal spreading of the chromosomes is important, with few chromosome overlapping and sufficient space between metaphase spreads and interphase nuclei (*Figure 2a*; see also *Plate 24a*). Following *Protocol 1*, about 100–200 slides with metaphase cell spreads are obtained. Before using these for a series of CGH experiments, samples from this batch should be tested by *in situ* hybridization with normal genomic DNA.

3 Isolation of genomic DNA

For optimal CGH experiments, a homogeneous painting pattern of the hybridized genomic probe is mandatory (see below). The quality of genomic probe DNA

Figure 1 (See *Plate 23*) Schematic illustration of CGH. Labelled genomic DNA from tissue samples to be analysed ('test DNA') as well as differently labelled DNA derived from normal cells ('control DNA') are hybridized simultaneously to normal metaphase chromosomes. As a result a general staining of all chromosomes is obtained (dashed lines). Chromosomal sequences present in additional copies within the test DNA (i.e. due to amplifications or polysomies) result in a higher staining at the corresponding chromosomal target sequences (indicated in black) as compared to control DNA. Losses of chromosomal sequences (i.e. due to deletions or monosomies) result in a weaker staining of the corresponding target chromosomes (indicated in a lighter staining). Measurements of fluorescence intensities along metaphase chromosomes by digital image analysis results in intensity ratio profiles (RP), which are plotted along chromosome ideograms (ID). Over- or underrepresentations of the corresponding chromosomal regions are indicated by ratio profiles exceeding defined cut-off levels.

critically influences the detection sensitivity of chromosomal imbalances by CGH experiments. If, for example, the incorporation of modified nucleotides by nick translation (see *Protocol 8*) is poor due to crude DNA preparations, this results in an irregular and speckled hybridization pattern of the metaphase chromosomes (*Figure 2e*; see also *Plate 24e*). As a result, differences in staining intensities are less pronounced. CGH can also be performed using DNA from paraffin-embedded tissue samples as well as DNA that has been amplified by universal PCR (see *Protocols* 6 and 7).

With regard to genomic control DNA, it should be noted that inter-individual differences in the staining of chromosomal regions containing high proportions of repetitive sequences can occur. It is desirable to obtain control DNA (e.g. from peripheral blood) from the same patient whose tumour DNA is examined. If this is not possible, it is advisable to prepare a batch of control DNA derived from one especially suitable donor, that can be used for a large number of CGH experiments. A number of chromosomal regions are known to be problematic for CGH analysis, most prominently 1p distal and chromosome 19. This problem can, in part, be overcome by repeating the experiment with the opposite labelling of the test and control DNA (experiment I: test DNA with fluorochrome I, control DNA with fluorochrome II); (experiment II: test DNA with fluorochrome II, control DNA with fluorochrome I) and comparing the ratio values of both experiments.

Figure 2 (See *Plate 24*) Examples of CGH experiments of sufficient and insufficient quality. (a) CGH experiment of good quality. A strong and homogeneous hybridization of the test and control DNA is visible. Some imbalances are already visible by eye (arrows). Note the good separation of the chromosomes. (b) Inhomogeneous illumination (strong in the upper part, low at the bottom) of the microscopic field due to insufficient adjustment of the optical path. (c) Average ratio profiles of chromosomes 3 and 4 with small 95% confidence interval indicating high quality of the CGH data. (d) Average ratio profiles of chromosomes 3 and 4 with wide 95% confidence interval due to weak and granular hybridization signals. (e) Example for granular hybridization signals after hybridization with genomic tumour DNA. (f) 'Flattening' of CGH average ratio profiles. Two hybridized chromosomes 8 from the same case to be averaged for the analysis are shown on the left and their individual ratio profiles are in the centre. Before averaging the ratio profiles are adjusted to predefined lengths. However, since chromosome condensation is not linear, this results in a shift of the ratio peaks, which indicate high level amplifications. Accordingly the averaged profile is flattened (see panel on the right). In extreme cases narrow amplifications and deletions remain undetected. (g) Example of average ratio profiles after CGH with DOP-PCR amplified DNA derived from ten cells. Unfixed cells resulted in correct CGH results, e.g. regarding the balanced state of chromosome 6 (left). Right: False positive results from CGH with ethanol/acidic acid fixed cells from the same cell line. (For explanations see text.) (h) CGH ratio profiles from an experiment performed with a single HL-60 cell according to the SCOMP protocol (*Protocol 7*). Note the high quality indicated by the narrow 95% confidence intervals. In this example, no false positive or false negative results were obtained.

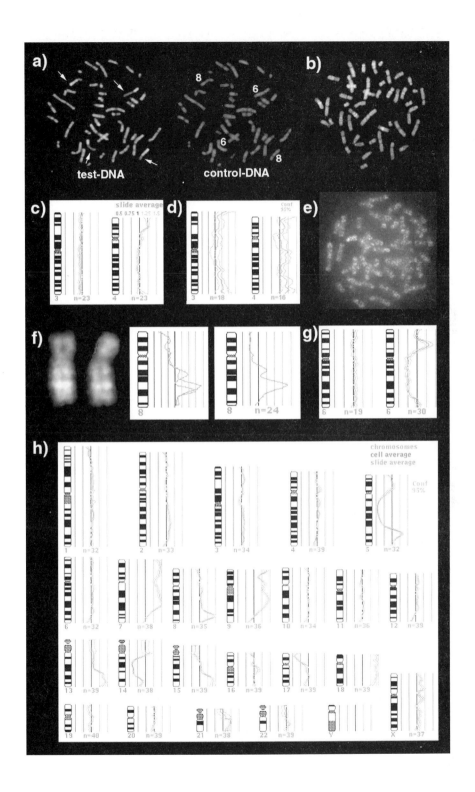

163

Protocol 1

Preparation of metaphase chromosomes from peripheral blood cells

Equipment and reagents

- 37 °C incubator
- One 50 ml polypropylene tube
- Cooled centrifuge (4 °C)
- Clean microscope slides pre-treated for cytogenetic preparations (see Chapter 3)
- 5 ml NH_4-heparinized blood
- Culture medium: RPMI 1640, 10% fetal calf serum, 1% L-glutamine, 1% penicillin and streptomycin, 1.5% phytohaemagglutinin, dissolved in water

- 10 µg/ml colcemid stock solution, stored at −20 °C
- Hypotonic solution: 3 g KCl, 0.2 g EGTA, 4.8 g Hepes, adjust to pH 7 with 1 M KOH
- Fixative solution: 1 vol. acetic acid, 1 vol. absolute methanol (ice-cold)
- 70%, 90%, 100% ethanol at room temperature

Method

1 Add 5 ml NH_4-heparinized blood to 50 ml culture medium.

2 Incubate at 37 °C in an atmosphere of 5% CO_2 for 72 h.

3 Add colcemid stock solution to a final concentration of 0.1 µg/ml. Incubate for 20 min at 37 °C.

4 Transfer to 50 ml tubes and spin at 200 g for 10 min.

5 Remove supernatant, resuspend cell pellet in remaining medium, and add a few drops of hypotonic solution.

6 Fill tube with hypotonic solution, carefully mix, and incubate for 15 min at 37 °C. Spin at 200 g for 10 min.

7 Remove supernatant, resuspend the pellet, and add a few drops of ice-cold fixative solution (prepare immediately before use and keep on ice).

8 Add slowly 5 ml of cold fixative solution in each tube while constantly whirling the suspension.

9 Spin at 200 g in a cooled (4 °C) centrifuge.

10 Repeat steps 7 and 8.

11 Incubate the tube on ice for 30–60 min.

12 Spin at 200 g in a cooled (4 °C) centrifuge and repeat steps 7–9 at least five times more (keep tubes on ice).

13 Resuspend cells in a small volume, e.g. 0.5–1 ml of fixative solution.

14 Drop the cell suspension from a distance of about 50 cm on pre-treated slides.[a,b] Slides should be kept in a moist environment during this procedure.

Protocol 1 continued

15 Keep the slides in a moist chamber for another 5 min, and then air dry at room temperature.

16 Dehydrate slides in a series of 70%, 90%, 100% ethanol for 5 min each.

17 Air dry again and keep slides at room temperature for one day before use.[c]

[a] Usually one drop of the cell suspension results in a sufficient amount of metaphase spreads for one CGH experiment. Check concentration of cells in the microscope. If the concentration of cells is too high, add fixative solution; if it is too low, spin again and resuspend in smaller volume. If large amounts of residual cell debris are still visible on the slides, repeat steps 9–11 several times.

[b] To clean microscopic slides wash in ethanol for several hours, dip several times in water, and dry immediately before cells are dropped.

[c] Slides can either be used directly for CGH experiments within a few days or sealed in containers and stored at –80 °C for several months. In order to prevent hydration, put slides in containers sealed in plastic bags with dryrite. Slowly thaw the slides before use.

Protocol 2

Isolation of genomic DNA from blood

Equipment and reagents

- Centrifuge tubes
- Pasteur pipette
- ~5–10 ml heparinized or citrate-buffered blood
- Lysis buffer: 155 mM NH_4Cl, 10 mM $KHCO_3$, 0.1 mM Na_2EDTA, adjust to pH 7.4 using HCl
- Chloroform/isoamyl alcohol (24:1)

- SE buffer: 75 mM NaCl, 25 mM Na_2EDTA, adjust to pH 8.0 using NaOH
- 3 M sodium acetate pH 5.2
- Isopropanol
- 70% ethanol
- TE buffer: 10 mM Tris, 1 mM Na_2EDTA, adjust to pH 8.0 using HCl

Method

1 Mix 1 vol. (about 5–10 ml) of heparinized or citrate-buffered blood and 3 vol. of lysis buffer. Shake carefully and incubate for 30 min on ice. Spin at 400 g for 10 min at 4 °C.

2 Discard supernatant and add 1 vol. of lysis buffer. Dissolve the pellet and spin at 400 g for 10 min at 4 °C.

3 Discard supernatant and add 0.5 vol. of SE buffer. Dissolve the pellet and spin at 400 g for 10 min at 4 °C. Discard supernatant.[a]

4 Add 0.5 vol. of SE buffer. Dissolve the pellet and add proteinase K to a final concentration of 100 µg/ml and sodium dodecyl sulfate (SDS) (final concentration of 1%). Mix carefully and incubate overnight at 37 °C in a water-bath.

5 Add 0.5 vol. of SE buffer and 1 vol. of equilibrated phenol/SE.[b]

Protocol 2 continued

6 Gently shake for 30 min and spin at 2000 g for 5 min at 10 °C. Transfer upper phase to a new centrifuge tube.

7 Add 0.5 vol. of equilibrated phenol/SE and 0.5 vol. chloroform/isoamyl alcohol. Shake for 30 min and spin at 2000 g for 5 min at 10 °C.

8 Transfer upper phase to a new centrifuge tube. Add 10 ml chloroform/isoamyl alcohol. Shake for 30 min and spin at 2000 g for 5 min at 10 °C.

9 Repeat step 8 two times.

10 Transfer upper phase to a new centrifuge tube. Add 0.1 vol. 3 M sodium acetate pH 5.2 and 1 vol. isopropanol.

11 Swirl the solution until a DNA precipitate is visible. Remove precipitate using a Pasteur pipette, the tip of which has been sealed and bent by heat. Briefly wash the DNA precipitate in 70% ethanol. Dissolve the DNA in TE buffer and store at 4 °C.[c]

[a] Sediment can be stored at this step at −70 °C.

[b] Phenol/SE. Melt approx. 200 g of crystalline phenol at 65 °C and add 0.25 g hydrochinolin. Add an equal volume of 1 M Tris–HCl pH 8.0 and agitate on a magnetic stirrer for 15 min. Remove the upper aqueous phase after separation from the lower organic phase. Add an equal volume SE buffer pH 8, stir 15 min, and again remove the aqueous phase after separation. Repeat this step three times but leave the aqueous phase after the last equilibration. Keep the phenol solution at 4 °C in a dark bottle for periods of up to one month.

[c] If no precipitate becomes visible (i.e. if only small amounts of DNA are prepared), isolate the precipitated DNA by centrifugation at 2000 g for 10 min. Discard the supernatant, wash the DNA pellet with 70% ethanol, and spin again at 2000 g for 10 min. Remove the supernatant, dry the pellet, and finally dissolve DNA in TE buffer.

Protocol 3

Isolation of genomic DNA from solid tissue samples

Equipment and reagents

- Razor blade
- Centrifuge tubes
- 10–100 mg cut tissue sample
- Extraction buffer: 10 mM Tris–HCl pH 8, 100 mM Na_2EDTA, 20 µg/ml pancreatic RNase (DNase-free), 0.5% SDS

- Proteinase K stock solution (to make a final concentration of 100 µg/ml in 20 mM Tris pH 7.0)
- Equilibrated phenol/TE[a]

Method

1 Cut tissue sample (about 10–100 mg) thoroughly into small pieces using a razor blade.

2 Incubate in 10 ml extraction buffer for 1 h at 37 °C.

3 Add proteinase K to a final concentration of 100 µg/ml and incubate for one to two days at 45 °C under constant shaking.

Protocol 3 continued

4 Add 10 ml of equilibrated phenol/TE. Gently shake for 1 h, spin at 2000 g for 5 min at 10 °C. Transfer upper phase to a new centrifuge tube.

5 Repeat step 4 twice more.

6 Proceed as described in *Protocol 2*, steps 8–11. Typical yields of DNA are approx. 100–1000 μg.

ª Phenol/TE, equilibrated to pH 7.5–8.0. Prepare as described in *Protocol 2*, but using TE instead of SE buffer.

Protocol 4

Isolation of genomic DNA from paraffin-embedded tissue

Reagents

- Xylol
- Methanol
- Extraction buffer: 25 mM EDTA, 75 mM NaCl, 0.5% Tween 20, 0.4 mg proteinase K at 55 °C

- 1 M sodium thiocyanate
- Proteinase K stock solution (to make a final concentration of 100 μg/ml in 20 mM Tris pH 7.0)

Method

1 Cut 30–50 slices from paraffin block, each 6 mm thick.

2 Remove paraffin by washing two times with 1 ml of xylol and two times with 1 ml of methanol.

3 Incubate overnight in 1 ml of 1 M sodium thiocyanate at 55 °C.

4 Digest for three days in 1 ml of extraction buffer at 55 °C. Add 0.2 mg of proteinase K each day.

5 Perform phenol extraction and proceed as described in *Protocol 2*, steps 5–11.

4 Isolation of single cells by micromanipulation

Tumour cells can be isolated from tissue sections (e.g. for the analysis of distinct tumour subregions) or from cell suspensions. Both methods have been used successfully in combination with CGH analysis (e.g. 7–9). While most frequently micromanipulators with glass needles (e.g. provided by Eppendorf) were used for such purposes, new laser-driven systems have been developed in the meantime. The current most popular devices are: the Laser MicroBeam Microdissection (LMM), the Laser Pressure Catapulting (LPC) system provided by P.A.L.M. Microlaser Technology, or the Laser Capture Microdissection (LCM) system (10). For these systems specialized protocols are provided by the suppliers. *Protocol 5* describes the isolation of individual cells by micromanipulation using glass needles.

Protocol 5

Micromanipulation of single cells

Equipment and reagents

- Petri dish
- 100 μm nylon filter (e.g. Falkon cell strainer)
- Cytospin
- Slides
- Micromanipulator equipped with extended glass needles
- Inverted microscope
- PCR tube
- PBS at 4 °C
- 3% paraformaldehyde in PBS
- Suitable specific antibody for detecting positive cells
- Digestion buffer (see *Protocol 6* or *7*)

Method

1. Prepare single cell suspensions by mechanically disaggregating tissue pieces. Scraping and cutting of samples in a Petri dish (optionally in PBS buffer) is followed by filtering cells through a 100 μm nylon filter.

2. Wash twice in PBS at 4 °C.

3. Fix cells for 20 min at room temperature in 3% paraformaldehyde (buffered in PBS) and wash three more times in PBS.

4. Adjust cell density to 1×10^6/ml to 5×10^6/ml. Cytospin 100 μl of the suspension at 100 g for 5 min on slides and air dry. Slides can be stored at this point sealed at –80 °C for several months.

5. Incubate with an appropriate antibody and detect positive cells or cell populations according to the recommendations of the supplier. As a final step wash slides in PBS or distilled H_2O and air dry.

6. For microdissection put slides under an inverted microscope. Cells should be well separated.

7. Isolate positively stained cells with a micromanipulator (e.g. from Eppendorf) equipped with extended glass needles. Try to put the needle below the cell to be isolated and peel cell off. (Depending on the cell type and the hydration status of the cells, this can be more or less difficult, and has to be optimized for every individual application.)

8. Break the tip of the glass needle to which the cell adheres and transfer it into a PCR tube with 20 μl of digestion buffer. Proceed as is described in *Protocol 6* or *7*, step 1.

5 Amplification of genomic DNA from small cell populations by universal polymerase chain reaction (PCR)

There are currently two universal PCR protocols that have been tested to be compatible with CGH analysis in more detail. Degenerate oligonucleotide primed

PCR (DOP-PCR) was described by Telenius *et al.* (11, 12) and is useful for the analysis of small cell populations (e.g. from defined tumour subregions). The second approach was described by Klein *et al.* (13) and termed SCOMP ('single cell comparative genomic hybridization'). In combination with CGH this method was shown to allow identification of chromosomal imbalances in single cells. Because of the interesting applications of these methods, they are both outlined in detail below.

It should be noted that PCR, in general, is a very sensitive approach and prone to artefacts by DNA contamination. In order to avoid this, only autoclaved and ultrapure materials and reagents have to be used and gloves should be worn. PCR reaction mixtures and PCR products have to be always prepared in separate rooms. Positive and negative controls should always be included in the experiments (see also Section 9; Troubleshooting).

Protocol 6

Amplification of small amounts of genomic DNA by DOP-PCR

Equipment and reagents

- 500 μl reaction tube
- Ovens at 50 °C and 95 °C
- PCR machine
- Isolated cells (20–200)
- 10 × DOP-PCR buffer: 2 mM MgCl$_2$, 50 mM KCl, 10 mM Tris–HCl pH 8.4, 0.1 mg/ml gelatine
- Proteinase K

- 10 × nucleotide mix: 2 mM of each dATP, dCTP, dGTP, dTTP
- 10 × DOP primer: 20 μM of the oligonucleotide 5′ CCG ACT CGA GNN NNN NAT GTG G 3′ (N = A, C, G, or T)
- Taq polymerase
- H$_2$O
- 1% agarose gel in TBE buffer

Method

1　Collect isolated cells (20–200) in a 500 μl reaction tube containing 20 μl DOP-PCR buffer to which 10 μg proteinase K is added. Incubate at 50 °C for at least 2 h.

2　Inactivate proteinase K by incubation at 95 °C for 20 min.

3　Add 8 μl of 10 × PCR buffer, 10 μl of 10 × nucleotide mix, 10 μl of 10 × DOP primer, 5 U Taq polymerase, and adjust to 100 μl with H$_2$O.

4　As a negative control combine the same solutions, but adding water instead of any DNA.

5　For universal amplification of the genomic DNA, the following temperatures and times are applied:

　(a) Initial denaturation of the DNA template, heat at 94 °C for 10 min.

　(b) 94 °C 1 min; 30 °C 1.5 min; 30–72 °C transition 3 min; 72 °C 3 min; for five cycles.

(c) 94 °C 1 min; 62 °C 1 min; 72 °C 3 min, with an addition of 1 sec/cycle; for 35 cycles.

(d) A final extension step at 72 °C for 10 min.

6 Run the PCR reaction on a 1% agarose gel in TBE buffer. A smear of DNA, ranging in size from about 200 bp to 2000 bp, should be visible. Compare to controls.

7 Label the amplified DNA by nick translation as described in *Protocol 8* and use as probe in a CGH experiment.

Protocol 7

Amplification of genomic DNA from single cells by SCOMP-PCR (13)[a]

Equipment and reagents

- 200 μl PCR reaction tube
- PCR machine at 42 °C with heated lid
- Ovens set to 37 °C, 65 °C, 68 °C, and 80 °C
- Sephadex columns
- Isolated cells (20–200 cells)
- Pick buffer: 50 mM Tris–HCl pH 8.3, 75 mM KCl, 3 mM $MgCl_2$, 137 mM NaCl
- 1 × proteinase K digestion buffer: 10 mM Tris–acetate pH 7.5, 10 mM Mg acetate, 50 mM K acetate, corresponding to 0.2 μl of 10 × One-Phor-All-buffer-Plus (Pharmacia), 0.67% Tween 20, 0.67% Igepal, 0.67 mg/ml proteinase K
- *Mse*I restriction enzyme
- *Mse*I buffer: 0.2 μl of 10 × One-Phor-All-Buffer-Plus, 0.5 μl *Mse*I (10 U), 1.3 μl H_2O
- 100 μM MseLig-21 primer: 5′ AGTGGGATTCCGCATGCTAGT 3′

- 100 μM MseLig-12 primer: 5′ TAACTAGCATGC 3′
- 10 × One-Phor-All-Buffer-Plus
- H_2O
- 10 mM ATP
- T4 DNA ligase (5 U; Roche Diagnostics)
- 10 × PCR buffer I (Roche Diagnostics 'Expand Long Template')
- 10 mM each dATP, dCTP, and dGTP
- 8.6 mM dTTP
- 1 mM biotin-16-dUTP (Roche Diagnostics Mannheim)
- 13 U of Thermo Sequenase
- Thermo Sequenase buffer: 260 mM Tris–HCl pH 9.5, 65 mM $MgCl_2$

Method

1 Transfer isolated single cell to a 200 μl PCR reaction tube containing 1 μl pick buffer.

2 Add 2 μl of 1 × proteinase K digestion buffer.

3 Incubate for 10 h at 42 °C in a PCR machine with a heated lid.

4 Inactivate proteinase K at 80 °C for 10 min.

5 For *Mse*I digestion add: 0.2 μl of 10 × One-Phor-All-Buffer-Plus, 0.5 μl *Mse*I (10 U), 1.3 μl H_2O, and incubate for 3 h at 37 °C.

6 Inactivate *Mse*I enzyme by heating the reaction mix for 10 min at 65 °C.

7 For ligation of oligonucleotides add: 0.5 μl of 100 μM MseLig-21 primer, 0.5 μl of 100 μM MseLig-12 primer, 0.5 μl of 10 × One-Phor-All-Buffer-Plus, and 1.5 μl H_2O.

8 Incubate in a PCR machine at 65 °C for 2 min, then slowly decrease to 15 °C with a ramp of 1 °C per min.

9 Add 1 μl of 10 mM ATP, 1 μl T4 DNA ligase (5 U), and incubate overnight.

10 Add to the 10 μl reaction mix: 3 μl of 10 × PCR buffer I, 2 μl of dNTPs (10 mM each dATP, dCTP, dGTP, and dTTP), and 35 μl H_2O.

11 Denature at 68 °C for 4 min to remove Mse-12 oligonucleotides.

12 Interrupt the PCR temperature program while keeping the tube at 68 °C. Add 1 μl (3.5 U) of DNA polymerase mixture of Taq and Pwo polymerase (Roche Diagnostics 'Expand Long Template'). Incubate at 68 °C for 3 min.

13 Start the following temperature program in the same PCR machine:
   (a) 94 °C 40 sec, 57 °C 30 sec, and 68 °C 1 min 15 sec, for 14 cycles.
   (b) 94 °C 40 sec, 57 °C 30 sec, and 68 °C 1 min 45 sec, for 34 cycles.
   (c) 94 °C 40 sec, 57 °C 30 sec, and 68 °C 5 min, for the final cycle.

14 For DNA labelling combine in a final volume of 30 μl:
   - 0.5 μl of LigMse-21 primer (100 μM)
   - 1 μl of dNTP (10 mM each dATP, dCTP, and dGTP; 8.6 mM dTTP)
   - 1.3 μl of 1 mM biotin-16-dUTP
   - 13 U of Thermo Sequenase
   - 3 μl Thermo Sequenase buffer
   - 0.5 μl of the primary PCR product (step 12)

15 Amplify reaction in a PCR machine with the following temperature program:
   (a) 94 °C 1 min, 65 °C 30 sec, and 72 °C 2 min, for 1 cycle.
   (b) 94 °C 40 sec, 65 °C 30 sec, and 72 °C 1 min, for 14 cycles.
   (c) 94 °C 40 sec, 65 °C 30 sec, and 72 °C 2 min, for 9 cycles.
   (d) 72 °C 5 min for the final cycle.

16 Remove ligated oligonucleotide by *Mse*I digestion.

17 Remove unbound nucleotides and digested oligonucleotides with Sephadex columns as described in *Protocol 8*, step 7.

[a] This protocol describes the amplification of genomic test DNA from a single cell for later CGH experiments. Regarding the control DNA for such experiments, optimal results were obtained by preparation of 0.5 μg of normal DNA in the same way as described in this protocol. Labelling is performed in step 14 with digoxigenin-11-dUTP instead of biotin-16-dUTP in this case.

6 Probe labelling

DNA probes for CGH can be labelled either directly or indirectly. Although directly labelled probes generate smooth fluorescence along the chromosomes, which is preferred for the analysis, indirect detection procedures also result in good quality CGH experiments. Commonly used haptens for indirect labelling are biotin, digoxigenin, and estradiol, and commonly used fluorochromes for direct or indirect labelling are FITC, rhodamine, or Texas Red.

DNA probes for CGH experiments are commonly labelled by the nick translation procedure which is based on the protocol of Langer *et al.* (14). Commercial labelling kits are available. Using varying DNase concentrations in the reaction solution, it is possible to adjust for an optimal probe size, which should be between 500–1000 nucleotides in length. The length of the probe molecules after labelling is a very critical factor for good quality *in situ* hybridizations (see above). In CGH it is important to adjust not only the length of the fragments, but also to achieve a similar length for test and control DNA. Alternatively PCR-based labelling procedures can be used, such as the SCOMP-PCR described in *Protocol 7*.

Protocol 8

Labelling of probe DNA by nick translation

Equipment and reagents

- 15°C incubator
- Boiling water-bath
- UV light tray
- 68°C oven
- 1 ml syringe packed with sialinized glass wool, and Sephadex G50 (in column buffer)
- 15 ml tube
- Centrifuge at room temperature
- Reaction tube
- 10 × reaction buffer: 0.5 M Tris–HCl pH 8.0, 50 mM MgCl$_2$, 0.5 mg/ml bovine serum albumin
- 10 × nucleotide stock for biotinylation of DNA: 0.5 mM dATP, 0.5 mM dCTP, 0.5 mM dGTP, 0.5 mM biotin-16-dUTP, 0.12 mM dTTP (alternatively, other biotinylated dUTP derivatives can be used)

- Probe DNA (2 μg)
- 10 × nucleotide stock for digoxigenin labelling of DNA: 0.5 mM dATP, 0.5 mM dCTP, 0.5 mM dGTP, 0.125 mM digoxigenin-11-dUTP, 0.375 mM dTTP
- 0.1 M 2-mercaptoethanol
- *E. coli* DNA polymerase I (20 U)
- DNase I (stock solution): 3 mg in 1 ml of 0.15 M NaCl, 50% glycerol
- Double distilled water
- 1–2% agarose gel in TBE buffer
- 0.5 μg/ml ethidium bromide
- 0.5 M EDTA
- 10% SDS
- Column buffer: 10 mM Tris–HCl pH 8.0, 1 mM EDTA, 0.1% SDS

Method

1 Combine for a 100 μl reaction:
   - 2 μg of probe DNA
   - 10 μl of 10 × reaction buffer

Protocol 8 continued

- 10 μl of 0.1 M 2-mercaptoethanol
- 10 μl of nucleotide stock (for biotinylation of DNA or for digoxigenin labelling of DNA)[a]
- 20 U of *E. coli* DNA polymerase I
- A tested volume of a 1:1000 dilution of DNase I
- Adjust to 100 μl with double distilled water (enzymes should be added last)

2 Incubate for 2 h at 15 °C.[b]

3 Put the reaction solution on ice. It should be kept on ice until the actual size of the reaction product has been determined.

4 Check the length of the probe molecules by gel electrophoresis. Take 10 μl out of the reaction solution, add gel loading buffer to the aliquot, and denature the solution by boiling in a water-bath for 2–3 min. After another 3 min on ice, the aliquot is loaded on a standard 1–2% agarose mini gel with a suitable size marker and run at 15 V/cm for 30 min. For the visualization of the DNA, stain the gel in 0.5 μg/ml ethidium bromide and take photographs under UV illumination.

5 For optimum hybridization conditions, the probe (which is visible as a smear) should be between 500 and 1000 nt in length. Depending on the result of the gel proceed as follows:

   (a) The probe size is within the desired range. Proceed to step 6.

   (b) The probe size is larger: add more DNase I, incubate at 15 °C (usually higher amounts of DNase I are added for another 30 min), repeat step 4.

   (c) The probe is not or almost not digested: purify probe and start the nick translation again.[c]

   (d) A considerable portion of the probe is below 100 nt in length: start reaction again using less DNase I.

6 In order to inactivate the DNase, 2 μl of 0.5 M EDTA (final concentration 10 mM) and 1 μl of 10% SDS (final concentration 0.1%) are added and the reaction mix is heated at 68 °C for 10 min.

7 Unincorporated nucleotides are separated from the labelled probe by gel filtration using a spin column. Since labelled probes exhibit a higher hydrophobicity than unlabelled DNA, presence of a detergent (e.g. 0.1% SDS) in the column buffer is mandatory.

   (a) Take a 1 ml syringe, pack it with sialinized glass wool up to the 0.2 ml mark. Pack the syringe with Sephadex G50 (in column buffer) to the 1 ml mark. Put the column in a 15 ml tube and spin it at 2000 g for 6 min at room temperature.

   (b) Remove flow-through, fill again, and repeat centrifugations until the resin is tightly packed to the 1 ml mark. Add 100 μl of column buffer and spin again at 2000 g for 6 min. Repeat this washing step three times. After the last washing step, make sure that the volume of the flow-through equals the volume of the loading buffer, i.e. 100 μl.

> **Protocol 8** continued

> (c) Prior to the centrifugation of the probe solution, put a small reaction tube in the 15 ml tube underneath the syringe. Load the probe solution on to the column and spin as before. The flow-through is collected in the reaction tube and contains the labelled probe at a concentration of approx. 20 ng/μl. It is ready for use in a CGH experiment or can be stored at –20 °C.
>
> Adequate spin columns are also commercially available.
>
> [a] Similarly, directly fluorochrome labelled nucleotides can be used instead of biotin-16-dUTP or digoxigenin-11-dUTP
>
> [b] When DOP-PCR products are used for labelling, digestion times need to be adjusted according to the length of the PCR products.
>
> [c] If DNase I has to be added for several times after the first digestion step, this is often due to protein contamination of the DNA. In these cases, an additional phenol extraction step should be performed (*Protocol 2*, step 5–11).

7 Comparative genomic hybridization

7.1 Denaturation of metaphase chromosomes

Denaturation of metaphase chromosomes is also a critical step in CGH. When the denaturation is not sufficient, no efficient hybridization can be performed. On the other hand, over-denatured chromosomes usually look fuzzy or may even be disintegrated. This also influences the identification of DAPI counterstained chromosomes which might become very tedious and time-consuming.

Protocol 9

Denaturation of chromosomal DNA on slides

Equipment and reagents

- Diamond pen
- 60 °C oven
- Slides
- Coplin jar in a 70 °C water-bath

- Denaturation solution: 70% deionized formamide, 2 × SSC, 50 mM sodium phosphate, adjust to pH 7.0 by adding HCl
- Ice-cold 70%, 90%, and 100% ethanol

Method

1 Select appropriate area for hybridization on the slide and mark it from underneath with a diamond pen.

2 Incubate slides in an oven at 60 °C prior to denaturation. This will prevent dropping of the temperature in the denaturation solution when the slides are added.

3 Fill a Coplin jar with denaturation solution and put it into a water-bath heated to 70 °C. Prior to denaturation, check temperature of the denaturation solution with a

thermometer inside the jar. This is a critical step!! For good results, the temperature should be at 70 °C.

4 Transfer the pre-warmed slides (not more than three at a time) into the Coplin jar with the denaturation solution for exactly 2 min.

5 Immediately transfer the slides into the Coplin jars with ice-cold 70%, 90%, and 100% ethanol for 5 min each.

6 After air drying, the slides are ready for hybridization.

7.2 Probe mixture

The large amounts of probe DNA provide high frequencies of labelled interspersed repetitive sequences to be suppressed. To accurately assess chromosomal imbalances by variations in the fluorescence intensity ratios, suppression must be as complete as possible (this is in contrast to FISH with single probes, whose localization is still clearly visible as long as they yield signals stronger than the background staining). Therefore an excess of unlabelled Cot-1 DNA is used and/or the pre-annealing time (after probe denaturation) is elongated.

Protocol 10

Probe mixture and denaturation

Equipment and reagents

- −70 °C freezer
- Eppendorf centrifuge at 4 °C
- Oven set at 37 °C and 75 °C
- Labelled test DNA
- Labelled control DNA
- 50–100 μg human Cot-1 DNA
- 3 M sodium acetate
- 70% and 100% ethanol
- Deionized formamide
- Vortex at room temperature
- Hybridization buffer: 4 × SSC, 20% dextran sulfate

Method

1 Combine 1 μg of each labelled test and control DNA and 50–100 μg of human Cot-1 DNA. Precipitate by adding 1/20 vol. of 3 M sodium acetate and 2 vol. of 100% ethanol. Mix well and incubate at −70 °C for 30 min.

2 Spin in an Eppendorf centrifuge at 2000 g for 10 min at 4 °C. Discard the supernatant, wash the pellet by adding 500 μl of 70% ethanol, and spin again (12 000 r.p.m., 10 min, 4 °C). Discard the supernatant, lyophilize.

3 Resuspend in 6 μl deionized formamide[a] by vortexing for >30 min at room temperature.

4 Add 6 μl of hybridization buffer[b] and again vortex for >30 min.

Protocol 10 continued

5 Denature DNA at 75 °C for 5 min, followed by incubation on ice for another 5 min, and pre-annealing of the probe solution at 37 °C for 20–30 min.

[a] Deionized formamide (molecular biology grade) is prepared using an ion exchange resin. The conductivity of the solution should be below 100 μSiemens.

[b] Prepare 20 × SSC and 50% dextran sulfate solutions. After careful dissolving, autoclave the dextran sulfate solution or filter it through a nitrocellulose filter. Mix 200 μl of 20 × SSC, 400 μl of 50% dextran sulfate, and 400 μl of double distilled water. Store at 4 °C until use.

7.3 *In situ* hybridization and signal detection

Protocol 11

In situ hybridization

Equipment and reagents

- 18 × 18 mm and 22 × 40 mm coverslips
- Wet chamber
- 37 °C oven
- Forceps
- Coplin jars
- Coplin jar wrapped in aluminium foil
- Shaking water-bath at 42 °C
- Hybridization mixture
- Pre-annealed probe (see *Protocol 10*)
- Denatured slide (see *Protocol 9*)
- Rubber cement
- Washing solution A: 50% formamide (inexpensive grade), 2 × SSC
- Washing solution B: 0.1 × SSC

- Washing solution C: 1% BSA, 4 × SSC, 0.1% Tween 20
- Washing solution D: 2 × SSC, 0.05% Tween 20
- Blocking solution: 3% BSA, 4 × SSC, 0.1% Tween 20
- Detection solution: 1% BSA, 4 × SSC, 0.1% Tween 20, 5 μg/ml fluorescein-conjugated avidin, 6 μg/ml rhodamine-conjugated anti-digoxigenin
- Counterstaining solution at room temperature: 2 × SSC, 200 ng/ml DAPI
- Antifade solution (e.g. VECTASHIELD, Vector Laboratories)

Method

1 Apply 12 μl of hybridization mixture with the denatured and pre-annealed probe on to the denatured chromosomes on the slides.

2 Put a 18 × 18 mm coverslip on top of the hybridization droplet. Take care not to trap air bubbles.

3 Seal the edges of the coverslip with rubber cement and put the slides in a wet chamber. Incubate for 48 h at 37 °C.

4 For signal detection pre-warm washing solutions A and B in a 42 °C and a 60 °C water-bath, respectively.

Protocol 11 continued

5   After taking the slides out of the wet chamber carefully remove the rubber cement using forceps.

6   Transfer slides into a Coplin jar containing pre-warmed washing solution A to 42 °C, agitate in a shaking water-bath for 10 min until coverslips come off. Transfer slides to another jar with washing solution A, agitate for 5 min. Change the washing solution A twice more, shake for 5 min each.

7   Transfer the slides to a Coplin jar containing pre-warmed (60 °C) washing solution B, wash for 5 min (42 °C water-bath). Change solution twice, each time washing for 5 min.

8   Take the slides out of the jar, drain them, and apply 200 µl of blocking solution.

9   Cover with a 22 × 40 mm coverslip, put slides in a wet chamber, and incubate at 37 °C for at least 30 min.

10 Let the coverslip slide off each slide, drain excess fluid, and add 200 µl of detection solution. Incubate in a wet chamber at 37 °C for 30 min. All subsequent steps should be carried out in dark Coplin jars (e.g. wrapped with aluminium foil).

11 Again, let the coverslip slide off, transfer the slides into washing solution C, wash 3 × 5 min at 42 °C (shaking water-bath).

12 Transfer slides into Coplin jar containing counterstaining solution and agitate at room temperature for 20 min.

13 Transfer slides into a jar with washing solution D, incubate for 1–2 min at room temperature.

14 Take each slide out of the Coplin jar, add 20–30 µl of antifade solution, and cover with a 22 × 40 mm coverslip. Put slides in suitable boxes, which should be kept at 4 °C for long-term storage.

8 Image acquisition and evaluation

Quantitative measurements of digital images in CGH experiments puts high demands on the equipment necessary for the evaluation, namely the microscope, the camera system, and the computer software program. The relevant instrumentation parameters specific for CGH are described elsewhere (e.g. ref. 15).

Image acquisition is performed using an epifluorescence microscope equipped with narrow bandpass filters selective for the fluorochromes used. It is very important that consecutive images from individual metaphase cells are obtained with no or only minimal optical shift. Settings with fixed multichroic and multiple bandpass emission filter combined with a moving excitation filter have turned out to be robust in this respect. New microscopic systems offer a motor-driven exchange or blocks containing fluorochrome adapted mirror and filter combinations. Due to the high accuracy of these systems optical shifts of the images are avoided.

Use of a field diaphragm in the excitation pathway is desirable in order to reduce the blurring and increase the contrast within the images. In our experience, a $63\times$ objective (UV transparent, plan) is most useful for evaluation by eye and image acquisition in CGH: metaphase spreads are sufficiently large for detailed analysis of small chromosomal subregions, yet small enough to fit in the 512×512 pixel margins required by most image analysis softwares currently available. Digital cameras should have high sensitivity, very homogeneous camera target response, and minimal shading.

There are several software programs commercially available which were designed for quantitative CGH image analysis. In general, these programs should fulfil several requirements. For easy handling, minimal user interactions and help with chromosome identification would be desirable. The program should be capable of segmentation of chromosomes in metaphase spreads. The majority of non-overlapping chromosomes of a given metaphase spread should be assessed in a quantitative evaluation. For an accurate diagnosis of over- or underrepresentations of chromosomal material, quality criteria for the selection of images and, even more importantly, for the rejection of inadequate images need to be included. The possibility to average values from several metaphase spreads is mandatory.

Regarding the cut-off levels given for diagnostic chromosomal over- and underrepresentations, there are different values described in the CGH literature. It is important to note that different threshold values were usually standardized empirically and depend on system parameters that are applied (e.g. the way background subtraction is performed). Therefore they should not be changed arbitrarily by the experimenter.

9 Troubleshooting of CGH experiments

For a good quality CGH experiment, a bright and homogeneous fluorescent painting of interphase cells and metaphase chromosomes should be visible after hybridization (*Figure 2a, c*; see also *Plate 24a, c*). Under these conditions gross aberrations, like gains or losses of a whole chromosome or distinct high-copy number amplifications, can already be detected visually without any equipment for digitized image acquisition or analysis (3). Fluorescence signals of the sex chromosomes can serve as an immediate visual quality feature. If control DNA hybridization of a male DNA does not result in a markedly weaker staining of the X chromosome(s), this indicates an inadequate quality of the hybridization experiment. Frequently this finding is accompanied by other hallmarks of poor quality, like a speckled, inhomogeneous staining pattern or high background fluorescence. Both may considerably disturb the quantitative fluorescence measurement by image analysis.

Some of the parameters that can account for poor hybridization results are:

(a) **The quality of metaphase cell spreads**. The preparation of adequate metaphase cell spreads is a prerequisite for high quality CGH results. We recommend to check every new batch of spreads for the criteria described in Section 3. If hybridization results are unsatisfactory, discard the whole batch and start a new slide preparation procedure. As shown in *Figure 2f* (see also *Plate 24f*), a strong variation of metaphase chromosomes regarding the condensation state leads to a 'stretching' of the average ratio profile resulting in an imprecise localization of chromosomal imbalances.

(b) **The quality of the DNA probe**. High amounts of protein in probe preparations often result in granular and inhomogeneous staining patterns (*Figure 2e*; see also *Plate 24e*). High granularity is also reflected by large confidence intervals (*Figure 2d*; see also *Plate 24d*), which indicate the variability of averaged ratio values. An improvement might be obtained by further purification of probe DNA (e.g. by proteinase K digestion and phenol extraction) and re-labelling.

(c) **Insufficient probe labelling**. Inappropriate size of labelled probe fragments after nick translation can also be the reason for granular or weak fluorescent signals: Too large fragments typically results in a starry background all over the slide; in contrast, if the probe size is too small, a homogeneous background distributed over all chromosomes can be seen. Therefore, checking the probe size after each labelling is highly recommended.

(d) **Insufficient suppression of repetitive DNA sequences**. Hybridization experiments can be severely impaired by insufficient suppression of repetitive sequences. This might become apparent by a strong staining of the heterochromatin blocks on human chromosomes 1, 9, and 16. An increase in the amount of the Cot-1 DNA fraction and longer pre-annealing times (the latter being less effective) can be used to achieve adequate suppression of signals in these regions of the genome.

(e) **Low quality of equipment**. The optical instrumentation, the camera system, and the image analysis software that is used can also contribute to insufficiencies. These are discussed elsewhere (see for example refs 4, 15, and 16). In *Figure 2b* (see also *Plate 24b*) an example of an inhomogeneous illumination of the microscopic field is shown. Signals of low intensity are prone to a higher rate of standard errors and therefore the corresponding chromosomes will be analysed with lower accuracy.

10 Troubleshooting of CGH experiments in combination with universal PCR

Using universal PCR, small populations or even single cells could be successfully analysed by CGH (see e.g. refs 7 and 13). It should be noted however, that this approach is susceptible to a number of artefacts. Most importantly this approach can lead to false positive and false negative results and therefore needs to be carefully controlled. Experiments with only a few cells containing known chromosomal aberrations revealed the occurrence of false positive and false negative results after DOP-PCR. Starting with 10–30 cells false negative results were primarily observed on smaller chromosomes. False positive results were preferentially observed as gains of chromosomal R-bands (e.g. on chromosomes 3p, 5q, 6p, and 12q) (*Figure 2g*; see also *Plate 24g*) and occurred with cells that were inadequately fixed or not sufficiently digested with proteinase K. It is wise to confirm, that amplified test and control DNA from normal cells, which have been fixed and enzymatically modified in the same way as the tumour cells, in fact indicate a balanced state in the subsequent CGH analysis. With small numbers of cells, there is also a risk of detecting presumed genomic imbalances, which are due to incomplete replication of the genome during S phase.

11 New developments: matrix-CGH

CGH has a limited resolution (several megabase pairs) due to the degree of DNA condensation in metaphase chromosomes and is technically demanding. Prior to the measurement of fluorescence intensity ratios, the chromosomes need to be identified prohibiting its application in high-throughput analyses. In the chip-based CGH approach termed 'matrix-CGH' (17) these restrictions are circumvented by substituting metaphase chromosomes as targets by well defined DNA fragments cloned in various vectors, which are immobilized on a glass surface (*Figure 3*; see also *Plate 25*) (17–19). Arrays are generated using commercially available robotic devices ranging from ink-jets to pin and capillary-based printing systems, which deliver droplets of DNA solution with a volume of less than one nanolitre. Although ink-jetting systems are fairly accurate, they are generally slow and appear therefore inferior to split pin systems for printing large scale microarrays. As for chromosome preparations, the humidity should be adjusted to facilitate delayed drying of droplets. Following the arraying procedure, the DNA is fixed and denatured prior to the comparative genomic hybridization, which is similar to *Protocols 8* and *9* (in addition to the fluorochromes mentioned in *Protocol 8*, Cy3 and Cy5 are a commonly used combination of dyes). However, one important modification of the protocol was invented to improve the hybridization efficiency (17): the hybridization temperature is increased to 52 °C to facilitate diffusion and the hybridization stringency is adapted by the formamide concentration adjusted to 50%. Digital images of the signals on the chip are acquired using either laser scanning devices or high aperture optics with a CCD camera. Fluorescence ratios are measured and calculated applying dedicated image analysis software. Criteria for the suitability of read-out devices include quality of the separation of the fluorescence signals of different wavelengths and sensitivity as well as linearity of the detection systems to allow accurate quantitative measurements of the fluorescence in both channels. Low copy number gains and losses are reliably scored with genomic DNA fragments in the size range of fragments cloned in cosmid, P1, PAC, or BAC vectors (17, 18), while high level amplification is detectable with targets only several kb in size. It should be noted that the use of several smaller fragments from one genomic region as targets also allows a reliable

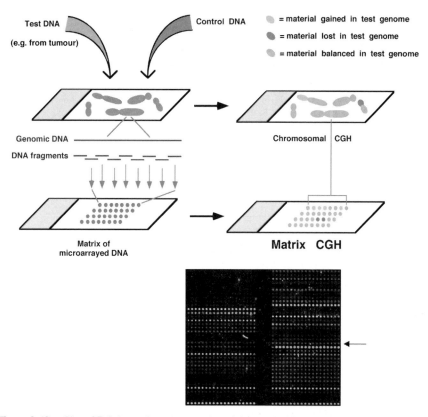

Figure 3 (See *Plate 25*) Schematic representation of the matrix-CGH approach, whereby chromosome targets are replaced by sets of well characterized genomic DNA fragments. Below the image of a chip is shown where target DNA fragments are arrayed in replicas of 20 and CGH allowed detection of a highly amplified gene.

scoring of low copy number differences by applying statistical means. Recently, an impressive resolution has been reported by the successful application of cDNA arrays for the assessment of DNA copy numbers (19). Matrix CGH DNA chips can be used to study microdeletions or overrepresentations and thus, represent powerful new tools for diagnostics and research in the fields of clinical and tumour genetics.

References

1. Kallioniemi, A., Kallioniemi, O.-P., Sudar, D., Rutovitz, D., Gray, J. W., Waldman, F., *et al.* (1992). *Science*, **258**, 818.
2. Du Manoir, S., Speicher, M. R., Joos, S., Schröck, E., Popp, S., Döhner, H., *et al.* (1993). *Hum. Genet.*, **90**, 590.
3. Joos, S., Scherthan, H., Speicher, M. R., Schlegel, J., Cremer, T., and Lichter, P. (1993). *Hum. Genet.*, **90**, 584.
4. Kallioniemi, O.-P., Kallioniemi, A., Piper, J., Isola, J., Waldman, F. M., Gray, J. W., *et al.* (1994). *Genes Chromosome Cancer*, **10**, 231.

5. Knuutila, S., Autio, K., and Aalto, Y. (2000). *Am. J. Pathol.*, **157**, 689.

6. Knuutila, S., Björkqvist, A.-M., Autio, K., Tarkkanen, M., Wolf, M., Monni, O., *et al.* (1997). *Am. J. Pathol.*, **152**, 1107.

7. Joos, S., Küpper, M., Ohl, S., von Bonin, F., Mechtersheimer, G., Bentz, M., *et al.* (2000). *Cancer Res.*, **60**, 549.

8. Weber, R. G., Scheer, M., Born, A., Joos, S., Cobbers, J. M. J. L., Hofele, C., *et al.* (1998). *Am. J. Pathol.*, **153**, 295.

9. Speicher, M. R., du Manoir, S., Schröck, E., Holtgreve-Grez, H., Schoell, B., Lengauer, C., *et al.* (1993). *Hum Mol Genet.*, **2**, 1907.

10. Emmert-Buck, M. R., Bonner, R. F., Smith, P. D., Chuaqui, R. F., Zhuang, Z., Goldstein, S. R., *et al.* (1996). *Science*, **274**, 998.

11. Telenius, H., Carter, N. P., Bebb, C. E., Nordenskjöld, M., Ponder, B. A. J., and Tunnacliffe, A. (1992). *Genomics*, **13**, 718.

12. Telenius, H., Pelmear, A., Tunnacliffe, A., Carter, N., Behmel, A., Ferguson-Smith, M., *et al.* (1992). *Genes Chromosome Cancer*, **4**, 257.

13. Klein, C. A., Schmidt-Kittler, O., Schardt, J. A., Pantel, K., Speicher, M. R., and Riethmuller, G. (1999). *Proc. Natl. Acad. Sci. USA*, **96**, 4494.

14. Langer, P. R., Waldrop, A. A., and Ward, D. C. (1981). *Proc. Natl. Acad. Sci. USA*, **78**, 6633.

15. Du Manoir, S., Kallioniemi, O.-P., Lichter, P., Piper, J., Benedetti, P. A., Carothers, A. D., *et al.* (1995). *Cytometry*, **19**, 4.

16. Piper, J., Rutovitz, D., Sudar, D., Kallioniemi, A., Kallioniemi, O.-P., Waldman, F. M., *et al.* (1995). *Cytometry*, **19**, 10.

17. Solinas-Toldo, S., Lampel, S., Stilgenbauer, S., Nickolenko, J., Benner, A., Döhner, H., *et al.* (1997). *Genes Chromosome Cancer*, **20**, 399.

18. Pinkel, D., Segraves, R., Sudar, D., Clark, S., Poole, I., Kowbel, D., *et al.* (1998). *Nature Genet.*, **20**, 207.

19. Pollack, J. R., Perou, C. M., Alizadeh, A. A., Eisen, M. B., Pergamenschikov, A., Williams, C. F., *et al.* (1999). *Nature Genet.*, **23**, 41.

Chapter 9
FISH in clinical cytogenetics

Jeremy A. Squire, P. Marrano, and E. Kolomietz
Ontario Cancer Institute, Princess Margaret Hospital, 610 University Avenue,
Toronto, Ontario M5G 2M9, Canada.

1 Introduction

In recent years the possibilities for visualizing several chromosomal targets simultaneously has meant that FISH analysis has an increasing role to play in the study of patient samples. The clinical cytogenetics service laboratory no longer focuses on the morphological analysis of chromosomal aberrations alone, but now has to provide comprehensive molecular cytogenetic analysis of diverse gene sequences consistently involved in certain classes of cancer and a variety of human genetic diseases (1). This increased clinical utility is due firstly to the diversity of DNA probes now available for the study of specific chromosomal changes in human cells, largely as a direct result of the success of the human genome project. Secondly, to the improvements in both labelling methods (see Chapter 2), and commercial imaging systems (see Appendix to this chapter), that have greatly facilitated the rapid transition of FISH from an esoteric research technique to one that fulfils a vital function in the clinical cytogenetics service laboratory. Finally, because valuable clinical information can often be obtained quickly by assessing the FISH signals present on interphase nuclei it is no longer always necessary to go through the lengthy process of preparing and analysing metaphase cells derived from patient samples.

Specific FISH probes and diagnostic protocols are now widely used for prenatal studies, dysmorphology, and as tumour-specific markers. Many new specific FISH clinical tests can assay for subtle chromosomal changes that would have been impossible to detect by conventional G-banded analysis. Target preparations for FISH analysis include metaphase and interphase cells derived from standard cytogenetic preparations as well as archived embedded histological material and fixed cytological preparations. This chapter will review the FISH methods commonly used in both cytogenetics and molecular pathology service laboratories, and will present protocols and troubleshooting tips of benefit to any investigator using FISH procedures with clinical samples.

2 Probes commonly used for FISH in the clinical laboratory

Probes utilized for FISH in the clinical laboratory are either generated in research laboratories or more usually they are purchased commercially (see Appendix for sources), and can be labelled directly or indirectly (see Chapter 2). Most commercial probes are supplied dissolved in hybridization solution and will already include appropriate blocking reagents. When using commercial probes it is essential that the supplier's instructions are followed closely, as each probe will have been optimized and tested using very specific stringency and hybridization parameters. Since the clinical applications of FISH are evolving rapidly it is to be expected that the commercial protocols and probe types used are being revised and improved regularly. We will therefore not reproduce some of the detailed protocols provided with commercial probes and kits, but will provide more general methods concerning sample preparation and analysis when using FISH in a service setting.

The easiest probes for the beginner are the chromosome-specific centromere probes, which allow the overall chromosome number or the sex chromosome constitution to be readily determined. Centromere probes usually generate intense signals because they hybridize to the highly repetitive DNA sequence motifs of the centromere. Useful probes for detecting terminal rearrangements are the telomere-specific probes. Whole chromosome-specific paint probes can be very helpful for confirming the identity of rearrangements in which G-banding has proved inconclusive. Increasingly, the use of SKY or M-FISH (see Chapter 10), is proving invaluable as a general screening method to detect chromosomal rearrangements. Similarly one company, Cytocell Inc. (see Appendix), has introduced a multiprobe device for FISH screening based on the attachment of combinations of chromosomal probes to the surface of a specially designed hybridization chamber. When hybridization solution is added to the multiple chambers within the chamber many separate FISH assays can be performed at the same time on a single slide.

To detect gene amplifications, translocations, or microdeletions, there are a number of commercial unique sequence gene probes derived from cosmids, BACs, or PACs (see Chapter 2), that often contain several hundred kb stretches of contiguous human genomic DNA. Some of the classes of commercial FISH probes available and their respective fluorescence intensity and general applications are presented in *Table 1*. At the end of this chapter, *Tables 2–4* detail the applications and assay methodologies currently used to address some of the clinical problems in which molecular cytogenetic testing is proving helpful. In the Appendix the web sites of the molecular cytogenetic companies that supply commercial FISH probes are provided.

Table 1 Different classes of commonly used probes used for FISH analysis in the clinical laboratory

Type	Fluorescence intensity	Typical applications
Centromere probes	Strong	Ideal probe for the beginner to learn FISH. Used for enumerating chromosomal monosomies, trisomies, and for determining the sex of nuclei for transplantation studies.
Telomere-specific probes	Moderate–weak	Used for determining whether small terminal rearrangements near chromosome telomeres have taken place.
Chromosome paints	Strong–moderate	Useful for identifying small marker chromosomes or aberrations when banding information is inconclusive (also see use of SKY/M-FISH paint probes in Chapter 10).
Translocation junction unique sequence probes	Moderate	For detecting the presence of chromosome translocations using metaphase and interphase cells.
Microdeletion unique sequence probes	Moderate	Identification of small submicroscopic deletions using metaphase preparations.
Probes to detect gene amplification	Strong	For detecting oncogene copy number increases (gene amplification) in interphase and metaphase cells.

3 Preparation of clinical samples for FISH analysis

Successful FISH depends largely on the initial quality of the patient sample. This presents a technical challenge for the clinical laboratory since the number and scope of applications of FISH is increasing as more commercial probes and hybridization kits become available. In addition, clinicians are becoming more aware of the significant advantages of performing FISH analyses on their patient's samples because of the speed, sensitivity, and specificity of FISH techniques. Thus FISH analysis is now being requested from diverse tissue types and results are usually required as fast as possible. In general the procedures described below will apply to cell types grown and prepared as suspension cultures and to cells grown on coverslips or slides. This section will review the various types of patient samples used for FISH and describe commonly used approaches for preparing optimal target slides for FISH in a clinical setting.

3.1 Preparation of metaphase chromosomes for FISH

Mitogen-stimulated lymphocytes from human peripheral blood are the most common source of human metaphases used for FISH in clinical cytogenetics. However, other mitotically active cells such as lymphoblasts, fibroblasts, fetal blood, amniocytes, and chorionic villus cultures, can also be used. For details of preparing metaphase chromosomes from human peripheral blood see Chapter 3, Section 3.1. Other texts comprehensively describe various procedures for optimizing the yield and quality of metaphase preparations derived from diverse cell types from both clinical genetics and cancer cytogenetics samples (1, 2).

3.1.1 Preparation of slides for FISH

Slide quality is one of the most important factors affecting hybridization efficiency and signal intensity. Slides should be evaluated using a phase-contrast microscope before hybridization. The cell density should not be too high and cells should be evenly distributed across the slide. Areas of cell clumping indicate poor slide preparation or sample quality. Under a phase-contrast microscope, nuclei and chromosomes should appear dark grey in colour (see Chapter 3, *Figure 1*). If they are light grey or very black and refractile they may not hybridize well. Visible cytoplasm surrounding metaphase spreads or interphase nuclei may also adversely affect hybridization and contribute to increased background. It is useful to apply a protease pre-treatment, as described in *Protocol 1*, when residual cytoplasmic material is detectable by phase-contrast microscopy. Most experienced cytogeneticists acknowledge that the quality of chromosomal preparations for both banding and FISH procedures can be greatly influenced by the drying environment during slide making. Humidity and temperature will significantly influence the rate of fixative evaporation during slide making. The drying time should be well controlled so that when the overall humidity is less than 30% slide making is performed near a steam source or humidifier. In recent years a controlled environmental evaporation chamber developed by Thermotron Inc. (see Appendix), has made for more reproducible slide making and has gained in popularity in clinical laboratories. In our experience the optimal humidity for drying slides for FISH is 55% and we prefer to dry slides at ~25 °C so that slides will take approximately 1–2 min to dry. A general protocol for performing routine FISH using metaphase preparations derived from clinical samples is presented in *Protocol 1*. Additional guidelines (as described for SKY/M-FISH) will also be generally applicable for troubleshooting clinical samples using any type of FISH probe (see Chapter 10, *Tables 1* and *2*).

3.1.2 Application of FISH procedures to cells grown on glass or plastic surfaces

Many clinical cytogenetics laboratories engaged in prenatal diagnosis use amniocytes or chorionic villi cultured on coverslips or glass slides. Solid tumours may also yield more representative cytogenetic preparations when cultured and harvested as attached colonies. To obtain sufficient cells for analysis, primary cultures must contain several large colonies of growth. It should be remembered that colony size and cell density will greatly influence the accessibility of the hypotonic treatment and fixative. Metaphase cells on the peripheral zone of colony growth are often more likely to have less cytoplasmic material and yield better quality FISH signals and hybridization efficiencies. If cytoplasmic material is apparent throughout a colony of interest when viewed by phase-contrast microscopy then the slide or coverslip should be treated with proteinase K (see *Protocol 1*, footnote a). Some of the theoretical considerations related to protease pre-treatments are discussed in Section 3.2.3.

3.1.3 Optimal age and storage requirements for FISH

In contrast to banding procedures where ageing for two to three days improves the morphology of the sample, FISH should be performed using fresher slides as discussed in detail in Chapter 4. In general, FISH can be performed the day following slide preparation with ageing slides at 37°C overnight in a dry incubator. However, from our experience, 3 h at 37°C is sufficient. This appears to improve both hybridization efficiency and signal brightness. In general, FISH and chromosome painting should be performed within two to three days of making slides. However, FISH will still yield reasonable results several weeks after slides are prepared, providing they are stored at room temperature in a dry atmosphere. Longer storage will lead to slow deterioration of slides. Some investigators feel that this deterioration in signal quality can be prevented if the slides are placed in dry airtight boxes sealed at −70°C for long-term storage. For indefinite storage of patient samples for future retrospective studies we prefer to store fixed cell pellets in 1.5 ml polypropylene cryotubes at −20°C. Cells stored in fixative should be rinsed in fresh fixative prior to making slides, as pH and water content may have changed during long-term storage.

3.1.4 Destaining previously G-banded samples for FISH studies

In some situations it may be necessary to perform FISH analysis on previously G-banded material. This procedure can be especially important for verifying an inconclusive chromosomal aberration on a specific metaphase cell. In this situation an image of the G-banded metaphase cell of interest should be captured and the microscope coordinates noted. Following the procedure described in *Protocol 1*, it is usually possible to obtain FISH results from samples that have been G-banded and stored as stained slides for one to three months. For older samples, slide storage variables such as humidity and ambient temperature will greatly influence success. It is thus very important to find out how the sample was prepared and stored. If there are a limited number of slides remaining from the patient sample of interest it is sometimes helpful to use a superfluous slide from another sample prepared the same week to derive the best conditions with respect to ageing and storage.

3.2 Preparation of interphase nuclei derived from clinical specimens for FISH

Very often only interphase nuclei are present in a clinical sample and special techniques are required to prepare and analyse such specimens for FISH procedures. The unique ability of interphase FISH analysis for documenting cell-to-cell heterogeneity makes interphase FISH particularly important in the analysis

Protocol 1

Performing FISH on metaphase spreads

Equipment and reagents

- Phase-contrast microscope
- For previously banded slides: one Coplin jar of xylene
- One Coplin jar of methanol
- One Coplin jar each of 70%, 80%, and 100% ethanol
- One Coplin jar of ice-cold 70% ethanol

- (Optional) protease digestion solution: 0.1 mg/ml proteinase K, 20 mM Tris pH 7.0, 2 mM CaCl$_2$
- 70% formamide/2 × SSC pH 7.0 pre-warmed at 72 °C in a water-bath
- Probe

Method

1 Prepare chromosome spreads as outlined in Chapter 3. For previously G-banded slides, slides should be treated with xylene for 5 min to remove residual oils, destained with methanol for 10 min, dehydrated in an ethanol series, air dried, and proceed to step 3. See Chapter 10, *Table 1*, for a general guide for preparing slides that give adequate G-banded chromosomes for subsequent use for FISH analyses.

2 Under phase-contrast microscopy, determine the extent of cytoplasmic residue on the slide preparation. If cytoplasm is present and pre-digestion with a protease is required use the protease digestion guidelines described.[a] If no pre-digestion is required proceed to step 3.

3 Denature the slides for 1.5–2 min in 70% formamide/2 × SSC at 72 °C. The time will vary according to the age and quality of the slide. For previously banded slides, a range of 20–30 sec is required.

4 Promptly place the slides into ice-cold 70% ethanol for 5 min following denaturation and proceed through an ethanol dehydration series, 80% and 100%, for 5 min each.

5 Prepare probe as instructed in supplier's instructions and perform FISH hybridization and wash protocol as described by supplier. The most common problems encountered can be high background or weak signals. Some general troubleshooting tips are described.[b,c,d]

[a] If protease digested is required, place the slides in a Coplin jar of proteinase K solution at room temperature for 7.5 min followed by dehydration of the slide in an ethanol series, 70%, 80%, and 100%, for 2 min each followed by air drying of the slide. Proceed with step 3 following the treatment.

[b] To reduce background soak off coverslip in 2 × SSC pH 7.0, 0.1% NP-40, rinse slide for 5 min at room temperature in fresh 2 × SSC pH 7.0, 0.1% NP-40, and proceed with the suppliers recommended final washing step with the temperature increased by 2 °C for 1 min.

[c] To improve signal intensity, the counterstain concentration may be decreased by destaining. Remove coverslip as described in footnote b, and immerse slide in fresh 2 × SSC pH 7.0, 0.1% NP-40 at room temperature for 10 min. Apply mounting medium and cover with a coverslip.

Protocol 1 continued

View with a fluorescence microscope. If the signal is still inadequate and background is not a problem then the hybridization may need to be repeated and the temperature of the high temperature wash should be decreased by 2 °C.

[d] The presence of multiple, bright star-like particles and speckling randomly distributed throughout the slide may be caused by disassociation or agglutination of the haptens or fluorophores conjugated to the DNA probe or detection reagents (see Chapter 2). Check the detection reagents by applying them to a control slide that was hybridized without probe. If this reagent is at fault it should be centrifuged and the supernatant can be used. If the probe is the cause of background then the supplier should be contacted. For non-commercial probes re-labelling and/or repeating ethanol precipitation may alleviate the problem (see Chapter 3, *Protocol 1*).

of mosaicism, tumour cell heterogeneity, and for monitoring allogeneic transplant engraftments. Similarly, interphase FISH can be very helpful for the analysis of small samples such as specimens obtained for pre-implantation embryo analyses, for prenatal diagnosis, from cytology smears, and from needle biopsies. Paraffin blocks are also widely used since most pathology laboratories routinely archive all formalin-fixed paraffin-embedded tissue for future retrospective studies. Readers are referred to the chapter by Hopman *et al.* (3) for more detailed discussion on some of the technical issues presented in this section.

3.2.1 Use of interphase nuclei derived from cytogenetic preparations

In all conventional metaphase preparations there are a large percentage of nuclei suitable for interphase FISH studies. The general guidelines described in *Protocol 1* apply equally to both metaphase and interphase nuclei. While no additional preparative technique is required for using such samples, there are important analytical and interpretative considerations regarding interphase FISH findings (see Section 4). As noted above, ageing and other procedures routinely used for preparing and storing conventional cytogenetic preparations may be detrimental for FISH procedures.

3.2.2 Use of interphase nuclei derived from cytological preparations

Routine blood films, cytological smears, cytospin, and tumour touch (or imprints) preparations will usually contain sufficient cells for interphase FISH studies. Often such preparations are 'dirtier' than cytogenetic preparations so additional pre-treatment steps such as proteinase K treatment (see *Protocol 1*, footnote a), are required to reduce background. For cytological preparations, cells on the slide should be fixed with a standard cytology fixative such as Shandon Cell-Fixx (Fisher), air dried, and then processed as described in *Protocol 1*, step 2.

3.2.3 Preparing tissue sections from paraffin-embedded material

Tissue samples such as solid tumours are usually fixed in formalin (a 37% commercially available formaldehyde solution) and then embedded in paraffin wax to preserve cell and tissue morphology for histopathological analysis. The success of FISH on paraffin-embedded specimens is directly dependent on the access-ibility of the target DNA within the cell nucleus and will be enhanced consider-ably by the utilization of pre-treatment methodologies (see *Protocol 2*), that increases the efficiency of the hybridization.

To prepare paraffin sections in the most optimum manner, specimens should be fixed as soon as possible upon receipt by the laboratory. Tissues are most commonly preserved in 10% formalin, other types of fixatives may be used, how-ever, certain agents alter DNA structure making the samples incompatible with FISH. Furthermore, if specimens are over-fixed the tissue will be dry and brittle, while under-fixation may produce soft and malleable tissue. Both of these characteristics render sectioning difficult and yield poor subsequent FISH results. A delay in tissue fixation or long incubations in a fixative solution will also lead to weak signals and reduced hybridization efficiencies.

Sections are usually prepared in the histology laboratory. Typically five sections at 5–8 μm on silanated slides are required and one haematoxylin and eosin (H&E) stained slide prepared. It helps to view the stained slide prior to the FISH experiment to determine if the section contains tumour or areas of interest. It is useful to get an experienced pathologist to look at the slide with you if it is difficult to interpret. A protocol for the preparation of paraffin sections is described in *Protocol 2*.

Once sections have been prepared the insoluble wax embedding compound must be removed. Deparaffinization is accomplished by immersing the slides in xylene followed by dehydration in ethanol to prepare the samples for pre-treatment. The pre-treatment procedure involves a series of incubations in sodium bisulfite (optional), a protein digestion solution, wash buffer, and an ethanol series in order to denature cellular proteins, aid in the removal of nuclear and extracellular matrix proteins, and remove protein crosslinks caused by formalin fixation. Digestion with a protease such as proteinase K or pepsin further enhances hybridization efficiency by digesting tissue proteins and making the target DNA more accessible for probe entry and subsequent hybrid-ization.

Troubleshooting suboptimal FISH results on paraffin sections may be difficult due to the large number of steps occurring between specimen procurement and FISH analysis. Proper preparation, embedding, and sectioning techniques are dependent on the type of tissue under investigation and these steps can intro-duce many variables to processing. Specimen variability will also introduce dif-ficulties relative to the tissue type and the amount of extracellular and cyto-plasmic material present. Certain tissues show greater resistance to protease digestion and prove difficult to analyse by FISH. Difficult samples may benefit from an increased incubation time in the protease solution or an increased incubation time in sodium bisulfite to increase the permeabilization of the tissue and to enhance the action of the enzyme (see *Protocol 2*, footnotes a and b).

Recently interphase FISH surveys of formalin-fixed tumour tissue have been greatly facilitated by the use of tissue arrays which contain multiple small circular punches derived from representative areas of a paraffin section of different tumours (4). The arrays can contain hundreds of different tumours represented as a 0.5–2 mm diameter disc of tissue containing several thousand cells which retain all the morphological features of the original specimen from which the tissue punch was obtained. Tissue arrays should be processed in much the same way as a regular paraffin section as described below.

Protocol 2

Preparation of paraffin sections for FISH

Equipment and reagents

- 45 °C water-bath
- Coverslips
- Fluorescence microscope with a FITC-propidium iodide filter set
- 90 °C and 37 °C oven
- Two Coplin jars containing xylene at room temperature
- Two Coplin jars containing 100% ethanol at room temperature
- Four Coplin jar of 2 × SSC pH 7.0 at room temperature

- Humidified slide chamber
- Ten 5–10 μm paraffin sections on silanated slides[a]
- Protein digestion solution at 45 °C: 0.5% pepsin in 0.85% NaCl pH 1.5
- Propidium iodide counterstain
- 70%, 80%, and 100% ethanol at room temperature
- Probe(s): see *Tables 2–4* and Appendices
- Rubber cement

Method

1 Add slides to a Coplin jar containing xylene. Gently agitate the slides at room temperature for 5 min. Transfer slides to another Coplin jar containing fresh xylene for an additional 5 min.

2 Transfer the slides to a Coplin jar containing 100% ethanol. Soak the slides for 5 min and then transfer slides to fresh 100% ethanol and soak for another 5 min. Agitate two to three times during each 5 min period.

3 Remove the slides from ethanol and allow to air dry.

4 Pre-warm protein digesting solution in a Coplin jar at 45 °C. Place slides in the Coplin jar of protein digesting solution and incubate at 45 °C for 15–20 min.[a]

5 Rinse the slide in 2 × SSC for 10–20 sec.

6 Apply 10 μl of propidium iodide or DAPI (depending on the label colour of the probe being analysed) counterstain to the slide and coverslip. View the slide using a FITC and propidium iodide filter set. Evaluate the tissue sections for under-digestion, appropriate digestion, or over-digestion.[b] For certain types of tissue additional troubleshooting steps may also be necessary.[c]

7 Rinse slides three times in 2 × SSC pH 7.0, agitating for 5–10 sec in each rinse.

8 Dehydrate slides in a series of ethanol washes, 70%, 80%, and 100% respectively for 1 min in each wash.

Protocol 2 continued

9 Allow the slides to air dry.

10 Prepare probe as described in supplier's instructions.

11 Place probe onto a coverslip and place slide onto probe/coverslip. Be careful to avoid air bubbles. Apply rubber cement along the perimeter of the coverslip. The coverslip must be well sealed to prevent evaporation of the probe mixture during denaturation.

12 Denature the probe and target DNA simultaneously by placing the slides in a 90°C oven for 12 min.

13 Transfer the slides to a pre-warmed humidified chamber, wrap with wet gauze, place in a sealable bag, and incubate at 37°C overnight.

14 Wash off the probe using stringency and washing conditions suggested by the supplier.

[a] The diameter of nuclei in various tissue types will vary between 4–8 μm. The estimated diameter for the tissue of interest should be ascertained from the literature and sections of the same thickness or 2–4 μm greater should be cut.

[b] At this stage it is possible to evaluate the degree of protein digestion before moving on to hybridization. A range of times is suggested because some tissues are more resistant to digestion than others and for some preparations proteinase K (footnote c), may work better than pepsin digestion. If the sample turns out to be under-digested it is possible to further treat the slides before hybridization. Over-digested tissue shows loss of cell borders. It will be difficult to see where one cell ends and another begins. If greatly over-digested, the cells may appear 'ghostly' or be lost altogether and digestion should be repeated with a new slide at a lesser time. Under-digested tissue shows persistent green autofluorescence and poor propidium iodide staining. Gently wipe off the immersion oil and remove coverslip by soaking the slide in $2 \times$ SSC pH 7.0 in a Coplin jar until the coverslip floats off. Transfer the slide to a Coplin jar containing fresh $2 \times$ SSC pH 7.0 for several minutes to clean off any residual oil. If further digestion is needed repeat step 4 (amount of time depends of how much digestion has already occurred). If no further digestion is needed proceed to step 7.

[c] Pepsin digestion works for most tissue types. Occasionally the tissue will exhibit persistent green autofluorescence of nuclei and no signal. Variability in histology fixation procedures necessitates variable time and/or concentration of pre-treatment. Certain tissue sections are more resistant to effects of protein digestion than others. This may be due to the tissue type or amount of protein crosslinking elicited by excessive formalin fixation. If no signal, very weak signal, and/or autofluorescence is obtained, it is possible to try pre-treating the slide with sodium bisulfite (see below) prior to digestion. This pre-treatment may be more necessary with normal tissue types than with tumour tissue. Also normal tissues are more refractory to protein digestion than tumour material derived from the same tissue type. If the protein digestion protocol outlined above does not remove autofluorescence and no signal is seen the following pre-treatment can be applied: pre-treat slide prior to protein digestion in 30% sodium bisulfite in $2 \times$ SSC pH 7.0 for 20 min at 45°C. If after pre-treatment and protein digestion there is still persistent green autofluorescence try using proteinase K as the digesting protein, i.e. proteinase K at 10 mg/ml in $2 \times$ SSC pH 7.0, at 37°C for 20 min. If there are a limited number of slides remaining from the patient sample of interest it is sometimes helpful to use a superfluous slide from another sample of the same age and cell type to derive the optimal digestion parameters with respect to ageing and storage.

3.2.4 Isolation of intact nuclei from paraffin-embedded tissue for FISH analysis

One of the drawbacks of using paraffin sections for FISH studies is that statistical methods (see Section 4.3.2), have to be employed to indirectly estimate whether numerical changes in chromosomal regions of interest have taken place. Statistical inference is not required if FISH is performed on intact nuclei that have been dissected from the region of interest on the section (reviewed in ref. 5). *Protocol 3* has proved helpful in our laboratory but does have the disadvantage that a large number of thick paraffin sections are required and the yield of nuclei can be poor with some tissue types.

Protocol 3

Isolation of intact nuclei from paraffin-embedded tissue for FISH

Equipment and reagents

- 37 °C oven
- 80 μm Nitex filter
- Microscope slides
- Cytospin (Shandon or equivalent) optional step
- Ten 50 μm paraffin sections
- 50 ml polypropylene centrifuge tubes
- Xylene
- 70%, 80%, and 100% ethanol
- PBS
- Proteinase K digestion solution: 10 mg/ml proteinase K, 0.5% SDS in TE

Method

1. Place paraffin sections into a 50 ml polypropylene centrifuge tube and add 20 ml of xylene. Gently agitate the tube for 10 min. Carefully decant off xylene and add fresh xylene for an additional 10 min with gentle agitation.

2. Replace xylene with 20 ml of ethanol and gently agitate for 10 min. Replace ethanol with fresh ethanol and agitate for an additional 10 min.

3. Rehydrate the tissue with 20 ml of 80% ethanol followed by 20 ml of 70% ethanol, 10 min each.

4. Replace ethanol with 50 ml water and centrifuge briefly to generate a loose pellet of tissue fragments. Carefully decant or aspirate as much of the water as possible without disturbing the pellet.

5. Add at least 5 ml of the proteinase K digestion solution,[a] and incubate overnight at 37 °C with gentle agitation.

6. Seat a Nitex filter on top of a new 50 ml centrifuge tube and add the tissue and solution from step 5 to the filter. Allow solution to pass through filter. Rinse the filter with 30–40 ml water. Remove the filter and spin the tube briefly to pellet cells.

Protocol 3 continued

7 Carefully decant or aspirate supernatant and resuspend single cell pellet in the residual volume.

8 If there is a very small pellet and few nuclei are suspected the nuclear pellet can be concentrated into one area of the microscope slide by centrifuging them with a Cytospin. If there appears to be a substantial number of nuclei in the pellet they can be dropped on microscope slides and air dried in the usual way.

[a] This volume should be ten-fold more than the tissue pellet volume after it has been decanted. The volume used should therefore be increased to ensure optimal enzyme activity.

4 Criteria for assessing and reporting FISH results

The minimum number of cells or metaphase spreads required to obtain a given result reflects the clinical context of the finding and the limitations of the patient material available for study. With tissue or cells that are hard to obtain, a single abnormal metaphase may be significant. For example in some situations limited FISH data may be supported by results obtained using PCR and/or Southern analysis.

Prior to enumerating or analysing FISH results on a patient sample it is important to carefully assess the overall quality, uniformity, and effectiveness of hybridization. Each hybridized slide should be evaluated for the specificity of the hybridization, the probe signal intensity, and the signal-to-background noise to determine if the hybridization was optimum for the given analyses. There should be minimal background or nuclear fluorescent 'noise'. At least 85% of all nuclei in the target area should be easily enumerable. For some applications (such as the detection of mosaicism or minimal residual disease) more rigorous analytical sensitivities and hybridization efficiencies are required (see Section 4.3.1).

Protocol 4

Pre-screening evaluation and determination of hybridization efficiency

Equipment and reagents

- FISH slides including patient slides and positive and negative controls

- Fluorescence microscope equipped with correct filter sets for probe set under study and a range of good quality fluorescence objectives. Analysis is performed using ×60 or ×100 objectives.

Method

1 Perform the FISH protocol appropriate for the type of patient and control slides in the experiment.

Protocol 4 continued

2 Prior to determining the hybridization efficiency, quickly scan the whole slide, noting the general signal-to-noise levels in different regions of the slide and any areas with high background or unusually weak signals. It may be useful to mark the underside surface of the slide in these areas with a diamond pen. If the signal intensity far exceeds background levels, proceed with estimating the hybridization efficiency.

3 Pick several representative areas of the slide and score at least 200 nuclei, following the selection criteria described in *Protocol 5*, from the areas selected and keep a running log of the number of cells scored and the observed signal counts for the patient and control slides.

4 For all slides add up the number of cells with no signal. In general, hybridization is considered to be adequate if greater than 85% of the cells scored have one or more signals. Lower hybridization efficiencies may be encountered with smaller probes (generated for example in a research laboratory). Extreme caution must be exercised when using probes with lower hybridization efficiencies or elevated background signal to provide clinical information.

4.1 General considerations when selecting cells for FISH microscopy

Generally look at all areas of the slide and analyse regions with uniformity in signal strength. Compare the intensity of the background signals to the intensity of the signals in the nuclei or metaphases of interest. The FISH signal intensity should be consistently greater than background intensity in the regions of the slide chosen for analysis. If the background signals are equivalent to signals in the nuclei then your counts will be skewed and the results biased.

Protocol 5

Selection criteria for FISH analytical microscopy

Equipment and reagents

• See *Protocol 4*

Method

1 Cells should ideally be analysed from all areas of the slide. Systematically select representative areas from different regions of the slide. Any regions that have unacceptable background or weak signals should have been identified in pre-screening evaluations (see *Protocol 4*).

2 Nuclei or metaphase cells should not touch or overlap.

Protocol 5 continued

3   Nuclei must have smooth well-rounded borders and the nuclei should be intact. Partially ruptured nuclear membranes may have lost informative chromatin. Metaphase spreads selected for study should similarly have no evidence of preparation artefacts or breakage.

4   Cells should not be surrounded by cytoplasmic material and there should be no evidence of potential drying artefacts such as rings or clumped cells.

5   Do not evaluate interphase nuclei with signals located on the extreme periphery of the nucleus.

6   Do not score regions of the slide containing nuclei that have no signals. Absence of signals may represent uneven or 'patchy' hybridization resulting in some areas of the slide having very weak or absent signals.

4.2 Scoring criteria for interphase FISH signal evaluation and enumeration

Interpretation of interphase FISH is very much dependent on statistical analyses and has inherent technical challenges. The presence of the signal is dependent upon the probe and its fluorescent label successfully entering the cell and hybridizing to the target DNA. The detection of the correct number of signals can be complicated by signals overlapping or splitting. Any background hybridization whatsoever will lead to major complications in interpretation. It is not uncommon to find 'monosomy' or 'trisomy' in nuclei that reflect technical artefact or 'false positive' background signals. Therefore, the accuracy of interphase FISH analysis is dependent upon recognizing these technical issues, correcting for them, and standardizing the scoring criteria accordingly.

Protocol 6

General guidelines for performing interphase FISH assays

Equipment and reagents

• See *Protocol 4*[a]

Method

1   Signals in a nucleus should in general have the same intensity.

2   It is usually necessary to focus up and down in the *z*-axis to accommodate spatial configurations of probe signals within the nucleus (for detailed discussion on three-dimensional considerations when performing FISH see Chapter 7).

3   Signals present in some nuclei that are more intense than the specific signal indicates the presence of regional background. Care must be taken when analysing any sample with this type of background noise.

Protocol 6 continued

4   Do not count lower level non-specific hybridization signals. These signals can usually be recognized by their lower intensity and different shape.

5   Some nuclei will have passed through the S phase of the cell cycle and may be present as G2 paired signals, i.e. two smaller signals in very close proximity. These paired signals represent a single chromosome already divided into chromatids and should be counted as one signal.

6   Count two signals connected by a strand of fluorescence as one signal. Sometimes centromere probes or long genomic probes,[b] will generate signals that are not spherical. As long as the signal is continuous it should be scored as one signal.

7   Count only nuclei in which a definite enumeration can be made, do not analyse or enumerate inconclusive cells.

8   Score 200 consecutive nuclei from each sample by two analysts,[a] so each scorer will analyse about 100 nuclei from a given sample.

[a] The slides should be coded and scored independently by two analysts. Any discrepancies may mean the established scoring criteria for the FISH assay are not being rigidly adhered to.

[b] Typically, FISH signals appear as separate fluorescent dots on each chromatid of a metaphase chromosome when the target size is 100–250 kb. Similarly in interphase nuclei such probes will also generate discrete easy to interpret signals. Larger probes can appear as fused signals straddling both chromatids and in interphase nuclei these probes can generate signals that present more diffuse or dispersed hybridization spots in the chromatin of interphase nuclei. Knowledge of the probe size and anticipated configuration in both metaphase spreads and interphase nuclei is essential.

4.3 Special considerations concerning interphase FISH interpretation

4.3.1 Analytical sensitivity of interphase FISH assays

Once a particular probe set is made available as a routine 'FISH test' as part of a clinical service it is important that the laboratory performs assay validation and establishes a database, which will allow reportable range and general laboratory experience with each probe to be established. The analytical sensitivity assay measures the success of a given FISH test in a particular laboratory environment and on a given tissue type. Since there are known differences in the cell populations and tissue types it is important to use the appropriate positive control tissue for the assessment of analytical sensitivity and for the establishment of the database. Analytical sensitivity analyses are performed by scoring 200 interphase nuclei representing at least five normal, preferably male, individuals (pooling of samples on one slide is acceptable). The nuclei are scored for the percentage of nuclei that exhibit the appropriate number of distinct signals. This assay will be performed in much the same way as described in *Protocol 5*.

For constitutional studies using FISH the recommended analytical sensitivity for probes intended for non-mosaic detection is 90% and for probes intended to detect mosaicism is 95%. Similarly, for detection of minimal residual disease sensitivities of 95% or greater are helpful. Databases for each probe should be established, it is then possible to determine the mean and standard deviation of results from a series of normal samples processed and analysed in the same manner as clinical samples. False positive rates can be then calculated and used for final scoring reports. More discussion on this issue is available in the following literature sources; URLs for VYSIS guidelines for single probe: http://www.vysis.com/tech_sup_fishproto_quality_single.asp and for dual probe: http://www.vysis.com/tech_sup_fishproto_quality_dual.asp; scoring criteria for pre-implantation genetic diagnosis of numerical abnormalities for chromosomes X, Y, 13, 16, 18, and 21 (6).

4.3.2 Statistical consideration concerning interphase FISH analysis of paraffin sections

Due to truncation of the nuclei during sectioning, loss of signal from areas of the nucleus excluded from the target slide will be encountered when enumerating signals after FISH has been performed on paraffin sections. The criteria for determining the significance of loss or gain of signals in interphase nuclei will depend on a number of parameters such as nuclear diameter, age of patient, type of tissue, etc. Readers are referred to some of the scientific literature where suggested cut-off values were adopted from the available literature (7). In our experience with FISH analysis of prostate cancer, chromosomal gains can been identified when more than ~10% of the nuclei exhibit more than two signals. Chromosomal losses have been identified when more than 50% of the nuclei exhibit a reduction of signal number and tetraploidy has been assumed when all chromosomes investigated show signal gains up to four.

5 Some of the commonly used FISH probes in clinical cytogenetics

5.1 FISH analysis of microdeletion syndromes

Microdeletion syndromes often present some difficulties in diagnostic determination. The usual FISH probe for a microdeletion syndrome is a two-colour probe mixture that includes a DNA sequence for the critical deleted region plus a control region reference probe used to confirm the identity of the target chromosome. These probes are designed for use on metaphase preparations. A normal metaphase cell would therefore show four signals when probing for an autosomal deletion syndrome, two of one colour for the control probe and two of the other colour for the region subject to deletion. An abnormal metaphase spread produces three signals, as the normal chromosome generates both a control signal and one of the other colour for the locus being tested, whereas on the deleted chromosome only the control signal will be seen. In *Table 2* some of the commonly encountered constitutional deletions are presented.

Table 2 FISH analysis of microdeletion syndromes

Microdeletion syndrome	Chromosomal location	Probe	Percentage detected with FISH
Prader–Willi	15q11-13	SNRPN	70% (8)
Angelman	15q11-13	D15S10	70% (8)
Miller–Dieker	17p13.3	LIS-1	90% (8)
Isolated lissencephaly sequence	17p13.3	LIS-1	33% (8)
DiGeorge	22q11.2	D22S75(N25) VYSIS: probe contains loci—TUPLE 1, D22S553, D22S609, and D22S942	85% (9)
Velocardiofacial (VCFS)	22q11.2	D22S75(N25) The same as for DiGeorge	53% (9)
Williams (Figure 1A, and Plate 26A)	7q11.23	LIM kinase Elastin gene	97% (10)
X-linked icthyosis	Xp22.3	Steriod sulfatase gene	85% (11)
Kallman	Xp22.3	KAL gene	Unknown (11)
Wolf–Hirschhorn	4p16.3	WHSCR 165Kb	Presumably >95% (12)
Cri du chat	5p15.2	D5S721	98% (13)

5.2 Use of the three-colour fusion (translocation/inversion) probes

If the probes based on green and red fluorescence used for FISH are close to specific translocation breakpoints on different chromosomes they will appear joined as a result of the translocation generating a 'yellow colour fusion' signal. In *Figure 1B* (see also *Plate 26B*) the detection of a Philadelphia chromosome in interphase nuclei of leukaemia cells is achieved by the presence of two double fusion (D-FISH) signals. All nuclei positive for the translocation contain one red signal (BCR gene), one green signal (ABL gene), and two intermediate fusion yellow signals because the 9;22 chromosome translocation generates two fusions, one on the 9q+ and a second on the 22q−. The following general guidelines may be helpful for performing this type of assay:

(a) A green and red signal that are juxtaposed but not overlapping should be scored as 'ambiguous'.

(b) Do not score nuclei that are missing a green or red signal. This assay is looking for the presence or absence of a fusion signal, not the absence of a green or red signal.

(c) Atypical signal patterns have been reported and these are now considered to be clinically important (14).

In *Table 3* some of the commonly used FISH assays in haematological cancer are presented. In addition to the scientific literature readers are referred to the

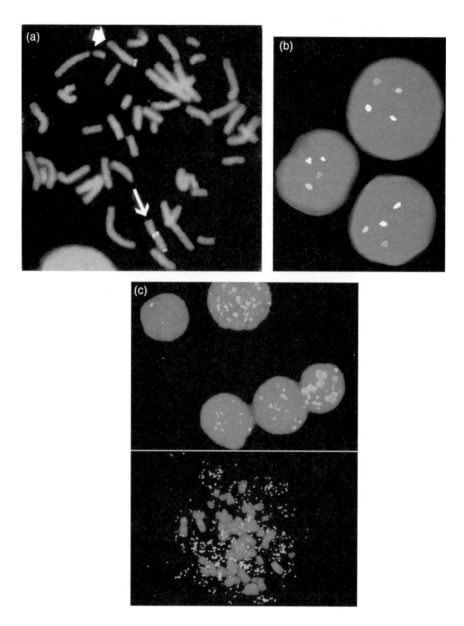

Figure 1 (See Plate 26) (A) Microdeletion analysis. The normal chromosome 7 (long arrow) shows yellow signals at both the telomeric control region and the Williams syndrome region at 7q11.23. The abnormal chromosome (short arrow) shows only the control signal, indicating a deletion at the Williams syndrome locus. (Photo courtesy of Dr Gordon DeWald, Mayo Clinic. Probes from Oncor, Inc.) (B) D-FISH. The Philadelphia chromosome is detected in interphase nuclei in leukaemia cells by the presence of two double fusion signals (yellow), along with the BCR gene (red) and the ABL gene (green). (C) MYCN FISH. Top panel shows red signals for the MYCN oncogene in a neuroblastoma. The cells show different levels of MYCN copy number showing the heterogeneity of the tumour. The lower panel shows a metaphase from the same patient with high levels of MYCN amplification and in the form of double minutes.

Table 3 FISH analysis of haematological malignancies

Neoplasm	Chromosomal location	Probe	Scoring method
CML/paediatric ALL	9q34/22q	*BCR/ABL*	Colour fusion observed in metaphase and interphase
Various leukaemias	11q23	*MLL*	Split signal, metaphase
Various leukaemias	5q31	EGFR1	Loss of signal, metaphase/interphase
Various leukaemias	7q31	DSS486	Loss of signal, metaphase/interphase
AML M4 EO	Inv(16)	CBFB	Split signal, metaphase
Various leukaemias	20q13.2	ZNF217, D20S183	Loss of signal, metaphase/interphase
Various haematological malignancies	8q24 /14q32	*MYCC/IgH*	Colour fusion observed in metaphase and interphase
AML-M1	12p13/21q22 8q22/21q22	*TEL/AML1 AML1/ETO*	Colour fusion observed in metaphase and interphase
AML-M3	15q22/17q21.1	*PML/RARA*	Colour fusion observed in metaphase and interphase
Various haematological malignancies, often when acceleration suspected	8cen	D8Z1	Enumeration using interphase and metaphase cells
Intersex transplant monitoring	Yp11.3/Xcen	SRY/DXZ1	Enumeration using interphase and metaphase cells

suppliers web sites, see Appendix, which will provide the most up-to-date listing of currently available probes and the preferred scoring method.

5.3 Use of FISH probes in assessing gene amplification

Gene amplification is one of the mechanisms by which cancer cells achieve overexpression of some classes of oncogenes and involves an increase in the relative number of copies per cell of a gene. This can range from one or two additional copies per cell to extreme examples where over a thousand copies per cell have been reported. Gene amplification can occur in association with the overexpression of oncogenes, thus conferring a selective growth advantage or as a mechanism of acquired resistance to chemotherapeutic agents and to poor prognosis. Gene amplification is highly suited to FISH analytical approaches which have the added benefit of excellent sensitivity and the ability to address cellular heterogeneity (15).

Neuroblastoma is characterized by the frequent occurrence of a highly amplified oncogene, MYCN. It has been known for many years that the presence of this aberration is strongly associated with poor outcome. More aggressive management is usually required when MYCN is found to be amplified. Similarly, breast cancer can be amplified for the oncogene HER2/Neu and presence or

Table 4 FISH analysis of solid tumours

Neoplasm	Chromosomal location	Probe	Scoring method
Neuroblastoma	2p24 2p23-24	MYCN LSI *N-myc*	Interphase, metaphase Amplification
Ewings sarcoma	11q24/22q12	FLI1/EWS	Colour fusion observed in metaphase and interphase
Rhabdomyosarcoma	2q35/ 13q14	PAX3/FKR	Colour fusion observed in metaphase and interphase
Breast cancer	17q11.2-q12	HER2/neu	Interphase amplification
Glioblastoma	7q?	EGF-R	Interphase amplification
Bladder cancer	Centromere regions of chromosomes 3, 7, 17, and 9p21 region of chromosome 9	CEP 3, 7, and 17 and LSI 9p21	Interphase enumeration

absence of this aberration may determine whether different treatment regimens are followed (16). Examples of metaphase and interphase FISH assays for gene amplification are shown in *Figure 1C* (see also *Plate 26C*). Some of the commonly detected aberrations observed in solid tumours which are amenable to FISH analysis are presented in *Table 4*.

6 Appendix (useful web sites for molecular cytogenetics clinical sources)

(a) Cytocell: http://www.cytocell.co.uk/

(b) Roche Molecular Biochemicals: http://www.biochem.roche.com/

(c) Thermotron: http://www.thermotron.com/cryogen.html

(d) Tissue array: http://www.tissue-array.com

(e) Ventana Medical Systems, Inc.: http://www.VENTANAMED.COM/

(f) Vysis: http://www.vysis.com/

References

1. Blancato, J. K. (1999). In *The principles of clinical cytogenetics* (ed. S. L. Gerson and M. B. Keagle), p. 443. Humana Press Inc., New Jersey.

2. Dracopoli, N. C. (2001). *Current protocols in human genetics*. John Wiley & Sons, Inc., New York, USA.

3. Hopman, A. H. N., Poddighe, P., Moesker, O., and Rameakers, F. C. S. (1992). In *Diagnostic molecular pathology: a practical approach*, (ed. C. S. Herrington and J. O'D. McGee), p. 142. IRL Press, Oxford, UK.

4. Kononen, J., Bubendorf, L., Kallioniemi, A., Barlund, M., Schraml, P., Leighton, S., *et al.* (1998). *Nature Med.*, **4**, 844.

5. van Lijnschoten, G., Albrechts, J., Vallinga, M., Hopman, A. H., Arends, J. W., Geraedts, J., *et al.* (1994). *Hum. Genet.*, **94**, 518.

6. Munne, S., Marquez, C., Magli, C., Morton, P., and Morrison, L. (1998). *Mol. Hum. Reprod.*, **4**, 863.

7. Qian, J., Bostwick, D. G., Takahashi, S., Borell, T. J., Brown, J. A., Lieber, M. M., *et al.* (1996). *Am. J. Pathol.*, **149**, 1193.

8. Ledbetter, D., Ballabio, A. (1995). In *The metabolic basis of inherited disease* (ed. C. Scriver, A. Beaudet, W. Sly, and D. Valle), 7th edn. p. 433. McGraw Hill, New York.

9. Crifasi, P. A., Michels, V. V., Driscoll, D. J., Jalal, S. M., and Dewald, G. W. (1995). *Mayo Clin. Proc.*, **70**, 1148.

10. Nickerson, E., Greenberg, F., Keating, M. T., McCaskill, C., and Shaffer, L. G. (1995). *Am. J. Hum. Genet.*, **56**, 1156.

11. Muroya, K., Ogata, T., Matsuo, N., Nagai, T., Franco, B., Ballabio, A., *et al.* (1996). *Am. J. Med. Genet.*, **64**, 583.

12. Wright, T. J., Ricke, D. O., Denison, K., Abmayr, S., Cotter, P. D., Hirschhorn, K., *et al.* (1997). *Hum. Mol. Genet.*, **6**, 317.

13. Pettenati, M. J., Hayworth, R., Cox, K., and Rao, P. N. (1994). *Clin. Genet.*, **45**, 17.

14. Sinclair, P. B., Nacheva, E. P., Leversha, M., Telford, N., Chang, J., Reid, A., *et al.* (2000). *Blood*, **95**, 738; Kolmietz, E., *et al.* (2001). *Blood*, **97**, 3581.

15. Obara, K., Yokoyama, M., Asano, G., and Tanaka, S. (2001). *Int. J. Oncol.*, **18**, 233; Squire, J. A., *et al.* (1996). *Mol. Diagn.*, **1**, 281.

16. Lebeau, A., Deimling, D., Kaltz, C., Sendelhofert, A., Iff, A., Luthardt, B., *et al.* (2001). *J. Clin. Oncol.*, **19**, 354.

Chapter 10

Multicolour FISH and spectral karyotyping

Jane Bayani and Jeremy A. Squire

Ontario Cancer Institute, Princess Margaret Hospital, 610 University Avenue, Toronto, Ontario M5G 2M9, Canada.

1 Introduction

It is often difficult to determine the identity of a small structural chromosomal aberration with certainty using conventional cytogenetic banding methods alone. The problems that can typically arise in both clinical and cancer cytogenetics are the presence of structural chromosome aberrations with unidentifiable chromosomal regions, or very complex chromosomes (sometimes called 'marker chromosomes') in which no recognizable region appears to be present. Confirmation of the cytogenetic origins of such chromosomal aberrations can sometimes be obtained by the judicious application of locus-specific FISH analysis (see Chapter 9) if the investigator has some general impression regarding a possible identity. However, such an approach is very subjective, risky, and requires some knowledge of the specific loci and available probes likely to be involved in the aberration. A more systematic approach is to use whole chromosomal paints in succession, until the marker chromosome and its constituents can be identified. While this strategy will eventually identify each chromosomal region involved, it is both costly and time-consuming and may lead to the depletion of valuable patient samples. Recently some generalized screening FISH techniques utilizing sensitive and differentially labelled chromosome-specific paints (see Chapter 2) have been developed, that allow the full chromosome complement to be analysed to identify unknown aberrations. In this chapter the commonly available methods will be described with suggested protocols and approaches to troubleshooting will be presented.

There are currently two or three slightly different imaging systems available that utilize the mechanical rotation of fluorescence excitation filters to distinguish the distinct fluorescence of a mixture of chromosomal paints during image acquisition. In this chapter such filter-based systems are generically termed multicolour FISH (M-FISH) (1, 2) (although each supplier has their own modified acronym for this technique). The second more frequently used system, called spectral karyotyping (SKY), utilizes image analysis based on Fourier transformation to spectrally analyse the differential fluorescence of each chromosomal paint (3). Both the SKY and M-FISH methods require the use of human whole chromosomal paints that are differentially labelled, so that each chromosome will have a unique combination of colours emitted following hybridization for identification purposes. Each method can be performed as a laboratory procedure and is capable of identifying the cytogenetic origins of all chromosomes in the complement in one image acquisition step.

At present the SKY system is probably better established and has resulted in a larger number of research publications based on the number of citations in PubMed (http://www.ncbi.nlm.nih.gov/PubMed/). The speed and accuracy of both SKY and M-FISH has been extremely useful in characterizing complex cancer cell lines (4–13), identifying novel translocations in sarcomas (14–17), carcinomas (18–23), leukaemias (24–28), and brain tumours (29–31). In addition, there have been numerous applications in clinical cytogenetics (32–39) and in murine cytogenetic analysis (40–43).

2 Spectral karyotyping (SKY)

SKY is a combination of optical microscopy, high resolution imaging, and the measurement of spectral emissions by Fourier spectroscopy (45) (*Figure 1A*; see also *Plate 27A*). The commercial SKY probes are derived from flow-sorted chromosomes that are amplified and labelled using DOP-PCR (44). A 24-chromosome probe cocktail is generated by the combinations of five pure dyes, namely rhodamine, Spectrum-Orange™, Cy5, Cy5.5, and Texas Red. This allows $2^n - 1$ or 31 combinations. Thus each chromosome has a unique spectral 'signature', generated by the specific combination of the five pure dyes. The generation of a spectral image is achieved by acquiring ~100 frames of the same image that differ from each other only by their optical path differences. Once a spectral image is acquired, the SKYView™ software compares the acquired spectral image against the combinatorial library containing the fluorochrome combinations for each chromosome and generates a 'classified' image. The classified image pseudo-colours the chromosomes to aid in the delineation of specific structural aberrations where the RGB (red–green–blue) display image (*Figure 1D*; see also

---

Figure 1 (See Plate 27) (A) Schematic representation of spectral karyotyping image acquisition and analysis. The hybridized specimen is visualized by fluorescence microscopy, where the image is passed through an interferometer. Differences in the optical path length are analysed by Fourier transformation and displayed as an interferogram. This data is then analysed by the SKYView™ software allowing for spectral identification of the fluorochromes hybridized to the chromosomes. (B) Schematic representation of filter-based M-FISH. The hybridized specimen is visualized by fluorescence microscopy where the specimen is acquired using a set of filters controlled by a computerized excitation filter wheel. For each filter, an image is acquired by a CCD camera. This information is then collected and passed through the M-FISH software. (C–E) Hybridization of an ovarian carcinoma metaphase spread using SKY. (C) Inverted DAPI analysis in which arrows indicate chromosomes that possess chromosomal aberrations identified by SKY. (D) Red–green–blue (RGB) image in which display colours reveal some of the abnormal chromosomes. (E) Classified SKY images of structural aberrations affecting chromosomes 1, 3, and 8.

206

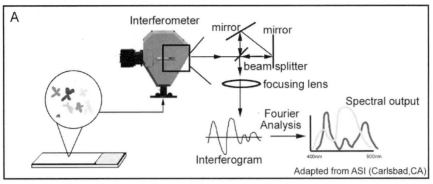

A

Interferometer

mirror mirror

beam splitter

focusing lens

Fourier
Analysis

Interferogram

Spectral output

400nm   900nm

Adapted from ASI (Carlsbad,CA)

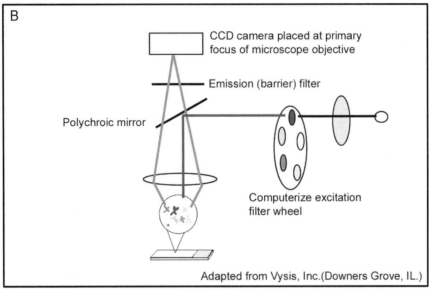

B

CCD camera placed at primary
focus of microscope objective

Emission (barrier) filter

Polychroic mirror

Computerize excitation
filter wheel

Adapted from Vysis, Inc.(Downers Grove, IL.)

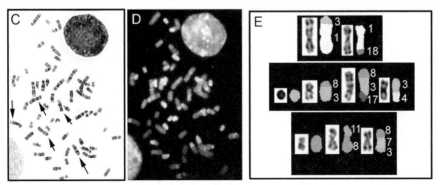

C

D

E

Plate 17D), which displays the fluorescent colours of the chromosomes may appear quite similar. For every chromosomal region, identity is determined by measuring the spectral emission at that point. Regions where sites for rearrangement or translocation between different chromosomes occur, are visualized by a change in the display colour at the point of transition. Examples of the use of SKY are illustrated in *Figure 1D, E* (see also *Plate 27D, E*).

3 M-FISH

M-FISH probes are generated in the same manner as the SKY probes, differing only by their dye combinations. Commercial M-FISH probes can be purchased from Vysis as SpectraVision™ probes. The fluorochromes used for this probe set includes SpectrumFar Red™, SpectrumRed™, SpectrumGold™, SpectrumGreen™, and SpectrumAqua™ used in a combinatorial fashion. Metasystems Inc. also provides M-FISH probes in their 24XCyte kit, which also has the added benefit of a 42-colour multicolour FISH (called mFISH) technique (armFISH), which permits the detection of chromosomal aberrations at the resolution of chromosome arms. The armFISH uses a commercially available mFISH reagent kit (24XCyte, MetaSystems GmbH) supplemented with a set of differentially labelled chromosome arm-specific painting probes (arm-kit, comprising either p- or q-arms of all human chromosomes, except the p-arm of the acrocentric and Y chromosomes) (2). All the M-FISH methods rely on the use of fluorochrome-specific filters moved sequentially during the image capturing process (1) (*Figure 1B*; see also *Plate 27B*). The identity of each chromosome or chromosome region is based on which combination of the five fluorochromes hybridizes together in one location. As with SKY, the unique identification dye combinations are stored in a reference library within the software to which the test specimen is compared and aberrations are delineated in the same manner as described for SKY.

4 M-FISH and SKY protocols

The methods described herein, can be divided into the following steps:

(a) Probe labelling.

(b) Metaphase spread pre-treatment and denaturation.

(c) Probe precipitation, denaturation, and hybridization to pre-treated slides.

(d) Post-hybridization washes and hapten detection.

Protocol 1

Probe labelling

1 See Chapter 2 for DOP-PCR generated SKY or M-FISH probes.

2 SKY probes are also commercially available from Applied Spectral Imaging in a kit containing all detection reagents.

3 M-FISH probes are also commercially available from Vysis Inc. or from Metasystems Inc.

Protocol 2

Spectral karyotyping metaphase spread pre-treatment and denaturation

Equipment and reagents

- 0.01 M HCl pre-warmed to 37 °C in a Coplin jar and maintained in an incubator or water-bath
- 10% pepsin (approx. 10–20 μl)
- Two Coplin jars containing 1 × PBS at room temperature
- One Coplin jar containing 1 × PBS/50 mM MgCl₂ at room temperature
- One Coplin jar containing 1%

- formaldehyde/1 × PBS/50 mM MgCl₂ at room temperature in a fume hood
- 70% formamide/2 × SSC pre-warmed at 72 °C in a water-bath
- One Coplin jar each of 70%, 80%, and 100% ethanol
- For previously banded slides: one Coplin jar of xylene, one Coplin jar of methanol

Method

1 Prepare chromosome spreads as outlined in Chapter 3. For previously G-banded slides, slides should be treated with xylene for 5 min to remove residual oils, destained with methanol for 10 min, and dehydrated in an ethanol series. No digestion is required; proceed to step 4. See *Table 1* for a general guide to preparing slides that give adequate G-banded chromosomes for subsequent use for SKY or M-FISH analysis.

2 Under phase-contrast microscopy, determine the extent of cytoplasmic residue on the slide preparation.

3 To the pre-warmed 0.01 M HCl, add 10–15 μl of 10% pepsin. Allow the slides to incubate for 5–10 min. The amount of time and pepsin may vary depending on batch variation of pepsin made as well as the quality of the chromosome preparation. It is therefore suggested that a range of concentrations and times be explored against representative chromosome preparations to determine the upper and lower thresholds.

4 Wash the slides in 1 × PBS for 5 min at room temperature. At this point, it is possible to view the slides again under phase-contrast microscopy to determine the extent of pepsin digestion. If required, the slides may be returned to the pepsin solution for further digestion.

5 Wash the slides in 1 × PBS/50 mM MgCl₂ for 5 min at room temperature.

6 Incubate the slides in 1% formaldehyde/1 × PBS/50 mM MgCl₂ for 10 min at room temperature in a well-ventilated area or fume hood.

7 Wash the slides in 1 × PBS for 5 min at room temperature.

8 Pass the slides through an ethanol dehydration series: 70%, 80%, 100% for 5 min each and allow the slides to air dry.

Protocol 2 continued

9 Denature the slides for 1.5–2 min in 70% formamide/2 × SSC at 72 °C. The time will vary according to the age and quality of the slide. For previously banded slides, a range of 20–30 sec is required.

10 Promptly place the slides into 70% ethanol following denaturation and proceed through the dehydration series as in step 8.

Table 1 General guide for processing cytogenetic preparations for G-banding analysis followed by SKY or M-FISH[a]

Banding steps	Slide 3 days –1 month	Slide >1 month	In situ slide preparation 3 days–1 month	In situ slide preparation >1 month
Trypsin (working solution)	10–20 sec with agitation	20–30 sec with agitation	40–60 sec with agitation	60–90 sec with agitation
Saline	Brief rinse	Brief rinse	Brief rinse	Brief rinse
Saline	Brief rinse	Brief rinse	Brief rinse	Brief rinse
Stain	50 sec	1+ min	50 sec	1+ min
Water	Brief rinse	Brief rinse	Brief rinse	Brief rinse
Water	Brief rinse	Brief rinse	Brief rinse	Brief rinse
Denaturing time	20–30 sec	45 sec–1 min	30–45 sec	1 min–?

[a] Adapted from Bayani et al. (46).

Protocol 3

M-FISH metaphase spread pre-treatment and denaturation

Equipment and reagents

- 1 M HCl pre-warmed to 37 °C in a Coplin jar and maintained in an incubator or water-bath
- 10 mg/ml RNase A stock solution
- 2 × SSC
- 10% pepsin stock
- One Coplin jar containing 2 × SSC at room temperature
- Two Coplin jars containing 1 × PBS at room temperature
- One Coplin jar containing 1% formaldehyde/1 × PBS/50 mM MgCl$_2$ at room temperature in a fume hood
- 70% formamide/2 × SSC pre-warmed at 72 °C in a water-bath

Method

1 Prepare chromosome spreads as outlined in Chapter 3. For previously G-banded slides, slides should be treated with xylene for 5 min to remove residual oils, destained with methanol for 10 min, and dehydrated in an ethanol series. No digestion is required; proceed to step 4. See *Table 1* for a general guide to preparing

Protocol 3 continued

slides that give adequate G-banded chromosomes for subsequent use for SKY or M-FISH analysis.

2 Under phase-contrast microscopy, determine the extent of cytoplasmic residue on the slide preparation.

3 Dehydrate the slides in an ethanol series for 5 min each and allow to air dry.

4 To 990 µl of 2 × SSC add 10 µl of 10 mg/ml RNase A and mix. Apply 20 µl of this solution to the slide and coverslip. Incubate at 37 °C for 30 min. Remove the coverslip and wash the slide in 2 × SSC for 5 min.

5 To the pre-warmed 1 M HCl, add sufficient 10% pepsin stock to reach a final 0.05% concentration. Allow the slides to incubate for 5–10 min. The amount of time and pepsin may vary depending on batch variation of pepsin made as well as the quality of the chromosome preparation. It is therefore suggested that a range of concentrations and times be explored against representative chromosome preparations to determine the upper and lower thresholds. Then wash the slides in 1 × PBS for 5 min.

6 Incubate the slides in 1% formaldehyde/1 × PBS/50 mM MgCl$_2$ for 10 min at room temperature in a well-ventilated area or fume hood.

7 Wash the slides in 1 × PBS for 5 min at room temperature.

8 Pass the slides through an ethanol dehydration series: 70%, 80%, 100% for 5 min each and allow the slides to air dry.

9 Denature the slides for 1.5–2 min in 70% formamide/2 × SSC at 72 °C.

10 Promptly place the slides into 70% ethanol following denaturation and proceed through the dehydration series as in step 8.

Protocol 4

Spectral karyotyping probe precipitation and hybridization to pre-treated denatured slides

Equipment and reagents

- Pre-treated, denatured chromosome slide preparations
- Glass coverslips
- Rubber cement
- Hybridization container
- 37 °C oven or incubator
- 75 °C water-bath or PCR machine
  For DOP-PCR amplified products:
- DOP-PCR labelled products
- Human Cot-1 DNA (1 mg/ml, Invitrogen)

- Salmon testes DNA (9.7 mg/ml, Sigma)
- 3 M NaAc
- 100% ethanol
- 70% ethanol
- 20% dextran sulfate/2 × SSC
- Deionized formamide
  Alternatively:
- SKYPaints™ (Applied Spectral Imaging, Carlsbad, CA)
- Vial 1 in the kit

Method

1   For DOP-PCR amplified product (one slide):

(a) Add 50 μl of human Cot-1 DNA for each 3 μl of amplified chromosome painting probes and 1 μl of salmon testes DNA.

(b) Add 0.1 vol. of 3 M NaAc.

(c) Add 2.5 vol. of 100% ethanol.

(d) Mix well and store at −20 °C overnight or at −80 °C for 1 h.

(e) Centrifuge at 2000 g in a refrigerated centrifuge for 20 min.

(e) Wash the pellet with cold 70% ethanol and repeat the centrifugation.

(f) Dry the pellet.

(g) Dissolve the pellet in 5 μl of formamide and incubate at 37 °C for 30 min with intermittent mixing.

(h) Add 5 μl of the 20% dextran sulfate/2 × SSC mix. Mix well and denature the probe at 75 °C for 7 min.

(i) Allow the probe to pre-anneal at 37 °C for 1 h.

2   Using commercially available SKYPaints™:

(a) For each slide, use 10 μl of the commercially available SKYPaints™ (ASI) for an 18–22 mm² coverslip.

(b) Heat denature the SKYPaints™ at 75 °C for 7 min.

(c) Allow the probe to pre-anneal at 37 °C for 1 h.

3   Following the annealing steps, use 10 μl of the pre-annealed probes (either DOP-PCR generated or commercially available probes) for a 18–22 mm² coverslip area and apply it to the denatured chromosome slide preparation.

4   Seal the coverslip with rubber cement and allow to hybridize in a container placed at 37 °C for 48 h. Previously G-banded slides need only be hybridized for 24 h.

Protocol 5

M-FISH probe precipitation and hybridization to pre-treated denatured slides

Equipment and reagents

- Vysis SpectraVision™ probe set
- Pre-treated chromosome slide preparations
- Glass coverslips
- Rubber cement
- Hybridization container
- 37 °C oven or incubator
- 75 °C water-bath or PCR machine

Protocol 5 continued

Method

1 For each slide, use 10 μl of the Vysis SpectraVision™ probe set and heat denature the probe for 5 min at 75 °C in a water-bath or PCR machine.

2 Apply 10 μl of denatured probe and coverslip, sealing with rubber cement.

3 Transfer the slides to a hybridization container and maintain at 37 °C for 24 h.

Protocol 6

Spectral karyotyping post-hybridization washes and hapten detection

Equipment and reagents

- Three Coplin jars of 50% formamide/ 2 × SSC pre-warmed to 45 °C
- Three Coplin jars of 1 × SSC pre-warmed to 45 °C
- Seven Coplin jars of 0.01% Tween 20/4 × SSC pre-warmed to 45 °C
- 70% ethanol, 80% ethanol, 100% ethanol (optional)
- One Coplin jar of 2 × SSC at room temperature
- One Coplin jar containing distilled water at room temperature
  For DOP-PCR generated probes:
- Blocking solution: 3% BSA/4 × SSC/0.01% Tween 20

- Avidin–Cy5 antibodies: diluted 1:200 in 1% BSA/4 × SSC/0.01% Tween 20 (Amersham)
- Mouse anti-digoxigenin: diluted 1:500 in 1% BSA/4 × SSC/0.01% Tween 20 (Sigma)
- Cy5.5 anti-mouse antibodies: diluted 1:200 in 1% BSA/4 × SSC/0.01% Tween 20 (Amersham)
- DAPI/antifade
  Alternatively:
- Commercially available kit from Applied Spectral Imaging
- Vial 2: blocking solution
- Vial 3: detection solution 1
- Vial 4: detection solution 2
- Vial 5: DAPI/antifade

Method

1 After hybridization, carefully remove rubber cement and ease off the coverslip, either by gently tapping or by immersing the slide in the first 50% formamide/2 × SSC solution and allowing the coverslip to fall off.

2 Wash the slides in 50% formamide/2 × SSC three times for 5 min each with gentle agitation.

3 Wash the slides in three washes of 1 × SSC for 5 min each with gentle agitation.

4 Dip the slides briefly in one wash of 0.01% Tween 20/4 × SSC.

5 Add sufficient blocking solution OR solution from Vial 2 to the slide and coverslip. Allow to incubate at 37 °C for 30 min.

Protocol 6 continued

6 Remove the coverslip and add sufficient diluted avidin–Cy5 antibodies and mouse anti-digoxigenin antibodies OR Vial 3 from the kit and coverslip. Return to 37 °C for 30–40 min.

7 Gently remove the coverslips and wash the slides in three washes of 0.01% Tween 20/4 × SSC for 5 min each. Gentle agitation may help to prevent high background.

8 Add sufficient diluted Cy5.5 anti-mouse antibodies to the slides OR solution from Vial 4 and coverslip. Incubate at 37 °C for 30–40 min.

9 Remove the coverslips and wash in three washes of 0.01% Tween 20/4 × SSC for 5 min each with gentle agitation.

10 Briefly rinse the slides in 2 × SSC.

11 Briefly rinse the slides in water.

An optional dehydration step may be carried out at this point.

12 Add sufficient DAPI/antifade and coverslip. The slides should be stored at −20 °C.

Protocol 7

M-FISH post-hybridization washes and hapten detection

Equipment and reagents

- One Coplin jar containing 0.4 × SSC/0.3% NP-40 at 73 °C
- One Coplin jar containing 2 × SSC/0.1% NP-40 at room temperature
- One Coplin jar containing 1 × PBS at room temperature
- DAPI/antifade (DAPI III, Vysis)

Method

1 Carefully remove the coverslip from the slide and immerse in 0.4 × SSC/0.3% NP-40 at 73 °C for 2 min.

2 Wash the slides in 2 × SSC/0.1% NP-40 at room temperature for 1 min.

3 Wash the slides in 1 × PBS at room temperature for 2 min.

4 Air dry the slides and stain with DAPI/antifade. Store slides at −20 °C.

5 General considerations for image acquisition and analysis

Criteria for successful image acquisition and analysis should be provided by the companies supplying the microscope, imaging equipment, and analytical software. However, some general points should be considered irrespective of which method of image acquisition and analysis is used. The overall signal intensity along the length of the chromosome should be high with minimal background.

The acquisition times, which are usually automatically set by the software, should be relatively short, as longer exposure times will amplify signals arising from background and increase the contribution of 'electronic noise' associated with all forms of digital imaging. Similarly, 'bright spots' which arise as a result of insufficient antibody or probe washing can skew automatic exposure times. This results in a low exposure of the metaphase of interest and increases the risk of ambiguous or erroneous chromosomal assignments.

Interpretation of the analysis relies both on the familiarity of the investigator with the basic technicalities of FISH assays, chromosomal identification by banding methods, and an understanding of fluorescence microscopy. One of the greatest concerns in the analysis is distinguishing between true translocation events and some of the preparation artefacts that may be misinterpreted during M-FISH and SKY. Regions of repetitive sequences such as centromeric areas and telomeric chromosomal ends can often give false classifications due to the differential suppression of those regions either with unlabelled Cot-1 DNA or by the labelled probe. Through careful analysis of normal metaphase spreads and confirmation by locus-specific FISH, real translocations can be distinguished from artefacts. Another phenomena which contributes to misclassification is the 'sandwich effect' where the junction between a true chromosomal translocation produces an intermediate colour which could be identified by the software as another chromosomal segment giving the false impression of a tiny insertion at the site of translocation. Since 24-colour painting analysis relies on the combination of up to five different fluorochromes, it is implicit that the fluorescence of two different chromosomal junction regions may in some circumstances be misclassified as a third chromosome undergoing insertion and bearing the spectra or dye combinations of the two partner chromosomes that are actually undergoing a simple rearrangement. This phenomenon can also be seen in chromosomes that are overlapping. The region of overlap will produce a colour pattern contributed by both chromosomes, and the software will identify the fluorescence at the region of overlap that most closely matches the reference library. In trying to determine whether a translocation is 'real' or not, it is advisable to look for longer chromosomes for analysis in another metaphase. If the 'insertion' is no longer present then it is likely arising due to suboptimal hybridization and/or the sandwich effect. Obviously it is necessary to confirm any uncertain rearrangement of interest using single whole chromosome painting or locus-specific FISH assays on another slide from the same preparation or by re-hybridizing following eluting chromosomal paints (see Chapter 9).

5.1 Image analysis using SKY

SKY is an interferometer-based method of image acquisition and analysis and has specific requirements for its correct operation. The ability for SKY to properly identify chromosomal segments relies on the proper calibration of the spectral reference library. Upon installation, the optical head is carefully aligned with the digital camera. Slides hybridized with each pure dye are then imaged to create the spectral reference library for that particular system and microscope. Since SKY analysis is based on the recognition of distinct spectral signatures, a less than optimal signal intensity can still generate a coherent chromosome classification, provided that there is no shift in the spectra. Furthermore, should a shift in spectra occur, it is possible to create a new spectral library based on the spectral signatures of hybridized slides. The advantage of SKY acquisition is the use of only the SKYCube™ to detect all the dyes and a short exposure using the DAPI filter for inverted banding images, rather than the lengthier exposure of the slide to five different filters as with M-FISH. The SKY software provides the capability to manually identify the spectral properties of each pixel along the length of a chromosomal region and compare this to the reference library. Such analyses can be very helpful in determining the identity of chromosomes in which identity is inconclusive.

5.2 Image analysis using M-FISH

M-FISH imaging relies on the sequential acquisition of images through specific filters and detecting the combination and the presence/absence of the reference dyes hybridization to the metaphase. This entails the acquisition of five different images that are then merged to form the final image. One of the most critical factors is the use of the proper filter sets designed specifically for the M-FISH probe cocktail being used. Careful consideration must be taken with respect to pixel shifts from one image to the next as well as the order in which the images are taken since it is known that some fluorochromes/dyes will emit a different wavelength upon degradation or exposure to UV. Such interaction between two fluorochromes could cause misclassification should, for example, a red fluorescing dye which degrades to a yellow emission after exposure to UV, be imaged before the yellow probe is exposed. Because the slides are exposed to different filters, there will be varying exposure times for each dye and this may also differentially quench the signals from the other fluorochromes/dyes later in the exposure sequence. Like the SKY software, most M-FISH software packages allow the user to identify the fluorochromes that are present along the chromosome of interest and comparing them to the combinatorial library. In the event of weak signals or small inconclusive regions of rearrangement this function is very convenient.

6 Troubleshooting

The success of multicolour painting approaches lies in the ability to identify regions of chromosomal transition as in the case of translocations, additions, and insertions. Furthermore, the criteria that underlie successful FISH experiments are as critical for SKY and M-FISH analysis. *Table 2* outlines some of the common problems encountered when carrying out 24-colour FISH analysis experiments with either system.

Table 2 Commonly encountered SKY and M-FISH difficulties

---

1. (a) Weak signal – slide

Slide age: Optimal results have been obtained from slides not older than two months in age. As the slides age, they become harder to denature. Conversely, very old preparations often have degraded DNA not adequate for FISH. Previously G-banded slides will have an even shorter lifespan and should be processed within two weeks.

Denaturation time: As the slide ages, the chromosomes become harder to separate into single strands. Slides used within one to two weeks of preparation should be denatured for 1.5–2 min. Slides that are older may require times that range from 2–3 min.

Cytoplasmic debris: The presence of cytoplasm may inhibit binding and contribute to background. A more aggressive protein pre-treatment may be required.

Excessive slide pre-treatment: Excessive enzymatic treatment may damage the target DNA making it less efficient for hybridization with the probe.

(b) Weak signal – probe

In-house probes: If probes are labelled in-house, strict controls must be undertaken to ensure that proper hapten or fluorochrome incorporation has been obtained to produce a high quality DNA probe.

Commercial probes: Usually the company has properly processed the product with the necessary quality controls. Check that the probes were properly stored and used within the expiration date.

Probe concentration: The probe is usually in excess of the target DNA, however, make sure that sufficient probe has been added to adequately cover the area of interest on the slide.

(c) Weak signal – technical

Sealing of coverslip: It is critical that the coverslip is adequately sealed, preventing any moisture from entering the hybridized area and diluting the probe.

Proper temperature: Check that the temperature in the oven or hotplate unit is correct for hybridization or incubations.

Incubation times: Avoid taking short cuts during incubation times with blocking reagents or detection reagents. Also be sure that the reagents do not dry up on the slide making washing more difficult.

Proper filter sets: For filter-based approaches, make sure that the proper filter sets are used for image acquisition. The use of improper filters can severely impair the ability to detect the correct signals or increase the amount of fluorescent cross-talk.

2. (a) Background – slide

Cytoplasmic debris: This is the most common culprit causing background. Increase the incubation time during the protein digestion or maintain the same time but change the concentration.

Bacterial/yeast contamination: Micro-organisms that have been cultured along with the specimen may have also made its way on to the slide. Unfortunately, this may be unavoidable.

Coverslips: If cells are to be grown *in situ*, use only glass coverslips. Plastic coverslips will cause autofluorescence.

Residual oils: Slides that have been previously visualized using immersion oil should be cleaned with xylene. Residual oils will prevent hybridization and cause background problems.

(b) Background – probe

In-house probes: Ensure that non-incorporated conjugated nucleotides are removed from the final probe preparation.

Table 2 (*Continued*)

Commercial probes: This is usually not the cause for background, especially when the use of the same probe on another slide preparation gives no background. However, note the lot number for future reference.

(c) Background – technical

Post-hybridization washes: Make sure that the correct temperature has been reached for the washes and incubations. Agitation during the washes can help to remove unbound probe and antibodies. Increasing the stringency of the washes by either increasing the temperature or altering the amount of SSC in the washes can also help. Avoid drying of the slide with any of the blocking or detection reagents.

Proper filter sets: As with weak signals, the use of improper filters to visualize and capture images may increase the cross-talk between fluorochromes that have similar emissions.

3. (a) Fading signals – slide
NA

(b) Fading signals – probe

In-house probes: Ensure that stocks of directly conjugated dNTPs are properly stored and used well within the expiry date. Unlike indirectly labelled probes, directly labelled DNA will have a much shorter shelf-life.

Commercial probes: Although these probes are supposed to be quality controlled, these too are under the same limitations of in-house probes. Take note of the lot number and expiry date of commercially made probes.

(c) Fading signals – technical

Antifade: Check that the DAPI/antifade solution is within its expiry date. Normally, the antifade is clear with a slight pink tinge. Antifade which has gone 'off' will turn increasingly amber in colour. Expired antifade medium will cause rapid signal degradation and display a red glow when viewed under the microscope.

Image acquisition: In the case of filter-based M-FISH, the sequence of image capture may influence the rate of signal degradation for subsequent dyes. Also long exposure under the DAPI filter may quench the Aqua signals.

4. (a) Poor chromosome morphology – slide

Over-denaturation: The cause of poor chromosome morphology is usually caused by over-denaturation causing the DNA to be destroyed. This can be sample-specific and/or slide age-related.

Previously G-banded slides

Over-processing: Over-trypsinization for optimal banding can cause the DNA to become sensitive to subsequent denaturation. For this reason, denaturation times are greatly reduced. This will also have an effect on the quality of hybridization and signal strength.

(b) Poor chromosome morphology – probe
NA

(c) Poor chromosome morphology – technical

Denaturation temperature: Check that the temperature for denaturation is accurate. The final internal temperature of the Coplin jar should be 72°C. Add 1° for each slide that will be denatured in that jar. Avoid denaturing more than five slides in one jar at one time. Plastic Coplin jars will have a greater differential temperature between the bath and the internal temperature, as compared to a glass Coplin jar.

General considerations include making sure that all solutions and labware are sufficiently cleaned, filtered, and/or sterilized. Items such as water or solutions containing BSA should be carefully checked for micro-organism or fungal growth. Keeping water-baths clean will help to prevent the addition of impurities (micro-organisms and other debris) that may result from condensation. Keeping antibodies and BSA-containing reagents refrigerated prevent bacterial growth. It is also suggested that antibodies be aliquoted and stored back at $-20\,^{\circ}C$ until ready for use. 'In-Use' aliquots can be stored for a few months at $4\,^{\circ}C$.

References

1. Speicher, M. R., Gwyn Ballard, S., Ward, D. C. (1996). *Nature Genet.*, **12**, 368.
2. Karhu, R., Ahlstedt-Soini, M., Bittner, M., Meltzer, P., Trent, J. M., and Isola, J. J. (2001). *Genes Chromosomes Cancer*, **30**, 105.
3. Schrock, E., du Manoir, S., Veldman, T., Schoell, B., Wienberg, J., Ferguson-Smith, M. A., *et al.* (1996). *Science*, **273**, 494.
4. Fan, Y. S., Siu, V. M., Jung, J. H., and Xu, J. (2000). *Genet. Test*, **4**, 9.
5. Allen, R. J., Smith, S. D., Moldwin, R. L., Lu, M. M., Giordano, L., Vignon, C., *et al.* (1998). *Leukemia*, **12**, 1119.
6. Bible, K. C., Boerner, S. A., Kirkland, K., Anderl, K. L., Bartelt, D. Jr., Svingen, P. A., *et al.* (2000). *Clin. Cancer Res.*, **6**, 661.
7. Liang, J. C., Ning, Y., Wang, R. Y., Padilla-Nash, H. M., Schrock, E., Soenksen, D., *et al.* (1999). *Cancer Genet. Cytogenet.*, **113**, 105.
8. Macoska, J. A., Beheshti, B., Rhim, J. S., Hukku, B., Lehr, J., Pienta, K. J., *et al.* (2000). *Cancer Genet. Cytogenet.*, **120**, 50.
9. Pandita, A., Zielenska, M., Thorner, P., Bayani, J., Godbout, R., Greenberg, M., *et al.* (1999). *Neoplasia*, **1**, 262.
10. Pan, Y., Lui, W. O., Nupponen, N., Larsson, C., Isola, J., Visakorpi, T., *et al.* (2001). *Genes Chromosomes Cancer*, **30**, 187.
11. Wong, N., Lai, P., Pang, E., Wai-Tong Leung, T., Wan-Yee Lau, J., and James Johnson, P. (2000). *Hepatology*, **32**, 1060.
12. Cottage, A., Dowen, S., Roberts, I., Pett, M., Coleman, N., and Stanley, M. (2001). *Genes Chromosomes Cancer*, **30**, 72.
13. Beheshti, B., Karaskova, J., Park, P. C., Squire, J. A., and Beatty, B. G. (2000). *Mol. Diagn.*, **5**, 23.
14. Carlotti, C. G. Jr., Drake, J. M., Hladky, J. P., Teshima, I., Becker, L. E., and Rutka, J. T. (1999). *Pediatr. Neurosurg.*, **31**, 307.
15. Joyama, S., Ueda, T., Shimizu, K., Kudawara, I., Mano, M., Funai, H., *et al.* (1999). *Cancer*, **86**, 1246.
16. Cohen, I. J., Issakov, J., Avigad, S., Stark, B., Meller, I., Zaizov, R., *et al.* (1997). *Lancet*, **350**, 1679.
17. Simons, J., Teshima, I., Zielenska, M., Edwards, V., Taylor, G., Squire, J., *et al.* (1999). *Am. J. Surg. Pathol.*, **23**, 982.
18. Adeyinka, A., Kytola, S., Mertens, F., Pandis, N., and Larsson, C. (2000). *Int. J. Mol. Med.*, **5**, 235.
19. Saunders, W. S., Shuster, M., Huang, X., Gharaibeh, B., Enyenihi, A. H., Petersen, I., *et al.* (2000). *Proc. Natl. Acad. Sci. USA*, **97**, 303.
20. Ghadimi, B. M., Sackett, D. L., Difilippantonio, M. J., Schrock, E., Neumann, T., Jauho, A., *et al.* (2000). *Genes Chromosomes Cancer*, **27**, 183.

21. Zitzelsberger, H., Lehmann, L., Hieber, L., Weier, H. U., Janish, C., Fung, J., *et al.* (1999). *Cancer Res.*, **59**, 135.

22. Speicher, M. R., Petersen, S., Uhrig, S., Jentsch, I., Fauth, C., Eils, R., *et al.* (2000). *Lab. Invest.*, **80**, 1031.

23. Dennis, T. R. and Stock, A. D. (1999). *Cancer Genet. Cytogenet.*, **113**, 134.

24. Veldman, T., Vignon, C., Schrock, E., Rowley, J. D., and Ried, T. (1997). *Nature Genet.*, **15**, 406.

25. Rowley, J. D. (2000). *Leukemia*, **14**, 513.

26. Rowley, J. D., Reshmi, S., Carlson, K., and Roulston, D. (1999). *Blood*, **93**, 2038.

27. Markovic, V. D., Bouman, D., Bayani, J., Al-Maghrabi, J., Kamel-Reid, S., and Squire, J. A. (2000). *Leukemia*, **14**, 1157.

28. Zhang, F. F., Murata-Collins, J. L., Gaytan, P., Forman, S. J., Kopecky, K. J., Willman, C. L., *et al.* (2000). *Genes Chromosomes Cancer*, **28**, 318.

29. Bayani, J., Zielenska, M., Marrano, P., Kwan Ng, Y., Taylor, M. D., Jay, V., *et al.* (2000). *J. Neurosurg.*, **93**, 437.

30. Bigner, S. H. and Schrock, E. (1997). *J. Neuropathol. Exp. Neurol.*, **56**, 1173.

31. Squire, J. A., Arab, S., Marrano, P., Bayani, J., Karaskova, J., Taylor, M. D., *et al.* (2001). *Mol. Diagn.*, **6**, 93.

32. De Krijger, R. R., Mooy, C. M., Van Hemel, J. O., Sulkers, E. J., Kros, J. M., Bartelings, M. M., *et al.* (1999). *Pediatr. Dev. Pathol.*, **2**, 577.

33. Krapp, M., Baschat, A. A., Gembruch, U., Gloeckner, K., Schwinger, E., and Reusche, E. (1999). *Prenat. Diagn.*, **19**, 610.

34. Marquez, C., Cohen, J., and Munne, S. (1998). *Cytogenet. Cell Genet.*, **81**, 254.

35. Ning, Y., Laundon, C. H., Schrock, E., Buchanan, P., and Ried, T. (1999). *Prenat. Diagn.*, **19**, 480.

36. Phelan, M. C., Blackburn, W., Rogers, R. C., Crawford, E. C., Cooley, N. R. Jr., Schrock, E., *et al.* (1998). *Prenat. Diagn.*, **18**, 1174.

37. Peschka, B., Leygraaf, J., Hansmann, D., Hansmann, M., Schrock, E., Ried, T., *et al.* (1999). *Prenat. Diagn.*, **19**, 1143.

38. Ramos, E. S., Rogatto, S. R., Marelli, L., Piram, A., Santos, S., and Squire, J. (In Press). *Hum. Genet.*,

39. Reddy, K. S., Sulcova, V., Young, H., Blancato, J. K., and Haddad, B. R. (1999). *Am. J. Med. Genet.*, **82**, 318.

40. Liyanage, M., Coleman, A., du Manoir, S., Veldman, T., McCormack, S., Dickson, R. B., *et al.* (1996). *Nature Genet.*, **14**, 312.

41. Shen, S. X., Weaver, Z., Xu, X., Li, C., Weinstein, M., Chen, L., *et al.* (1998). *Oncogene*, **17**, 3115.

42. Coleman, A. E., Forest, S. T., McNeil, N., Kovalchuk, A. L., Ried, T., and Janz, S. (1999). *Leukemia*, **13**, 1592.

43. Wiener, F., Kuschak, T. I., Ohno, S., and Mai, S. (1999). *Proc. Natl. Acad. Sci. USA*, **96**, 13967.

44. Telenius, H., Pelmear, A. H., Tunnacliffe, A., Carter, N. P., Behmel, A., Ferguson-Smith, M. A., *et al.* (1992). *Genes Chromosomes Cancer*, **4**, 257.

45. Malik, Z., Dishi, M., and Garini, Y. (1996). *Photochem. Photobiol.*, **63**, 608.

46. Bayani, J., Pandita, A., and Squire, J. A. (2000). *Elsevier Trends Journal Technical Tips Online* http://www.biomednet.com/db/tto

Chapter 11

cDNA microarrays for fluorescent hybridization analysis of gene expression

Javed Khan[1,2], Lao H. Saal[2], Michael L. Bittner[2], Yuan Jiang[2], Gerald C. Gooden[2], Arthur A. Glatfelter[2], and Paul S. Meltzer[2]

[1]Oncogenomics Section, Pediatric Oncology Branch Center for Cancer Research, National Cancer Institute, National Institutes of Health Advanced Technology Center, Room 134E, 8717 Grovemont Circle, Gaithersburg, MD 20877, USA

[2]Cancer Genetics Branch, National Human Genome Research Institute, National Institutes of Health, Building 49, Room 4A-15, 49 Convent Drive, MSC 4470, Bethesda, Maryland 20892-4470, USA

1 Introduction

As a consequence of the human genome project (HGP) the sequence of the entire human genome will be completed by the year 2003 (1). It is estimated that the 3 billion base pairs in the human genome code for between 28 000–1 200 000 genes (2–6). Perhaps 50–90% of these genes are represented by the ~1 800 000 human expressed sequence tag (EST) clones that are publicly available (http://www.ncbi.nlm.nih.gov/dbEST/dbEST_summary.html). These sequenced EST clones are invaluable research reagents that have allowed the emergence of three new techniques that permit the investigation of genome-wide expression analysis of any source of RNA. Included in this list are cDNA microarrays, serial analysis of gene expression (SAGE), and oligonucleotide probe array chips. It is helpful to briefly mention the alternative strategies to the cDNA microarray for gene expression analysis.

1.1 Serial analysis of gene expression

The technique of SAGE, first described in 1995 (7), utilizes high-throughput sequencing technology to obtain a quantitative profile of gene expression. In this method short nine to eleven base nucleotide sequence 'tags' from a test sample RNA are manufactured, concatenated, cloned, and sequenced. The expression level of a gene is thus measured indirectly by quantifying the number of tags which represents a particular gene. By this method it is possible to determine gene expression pattern characteristics of that sample as well as the identification of new tags corresponding to novel transcripts. Although SAGE has the potential to generate genome scale expression profiles, human cDNA micro-arrays have the advantage that they are readily amenable to the analysis of multiple samples thereby generating a large amount of gene expression data for statistical analysis.

1.2 Oligonucleotide arrays

In this method oligonucleotides (25-mers) are manufactured on to a glass surface by a light-directed chemical synthesis process using a combination of solid phase chemical synthesis with photolithographic fabrication techniques employed in the semiconductor industry (8). A more recent modification has been the use of a modified ink-jet printing process to produce oligonucleotides on a glass surface using phosphoramidite chemistry (see http://www.rii.com/tech/inkjet.htm). Like cDNA miocroarrays, probe arrays allow global gene expression of thousands of genes at once but has the disadvantage of the necessity for a priori knowledge of the gene sequence.

1.3 cDNA microarrays

Gene expression monitoring using microarrays was first described in 1994 by Drmanac *et al.* (9, 10) who used radioactive targets hybridized onto filter-immobilized PCR-amplified cDNA probes. The principle of cDNA microarray is akin to reverse Northern hybridizations where multiple probes in the form of cDNA clones are immobilized on a solid impermeable substrate such as glass and interrogate a labelled RNA population. Like Northerns cDNA microarray hybrid-ization takes the advantage of the property of DNA to form duplex structures between two complementary strands. Unlike Northerns and the other DNA hybridization-based techniques including, dot-blot, Southern blot, and FISH (fluorescence *in situ* hybridization), microarrays take advantage of being high-throughput due to automation allowing expression analysis of several thousand genes in one hybridization.

The two-colour fluorescence detection scheme has significant advantages over radioactively labelled hybridization by allowing rapid and simultaneous differential expression analysis of independent biological samples. Also use of ratio measurements compensates for target to target variations of intensity due to DNA concentrations and hybridization efficiencies.

This technique was first described by Schena *et al.* (11) who printed 48 genes of *Arabidopsis thaliana* on to glass slides and measured differential expression of genes between two different tissues, root and leaf. Fluorescent probes were made from each of these tissues by reverse transcription of mRNA using distinct fluorochromes. By measuring intensity ratios for each printed gene, they were able to show widespread differences in gene expression between these two tissues.

The cDNA microarray technology is rapidly evolving with constant improve-ments being made to both equipment and protocols, however we will describe to the reader the current methodology as outlined below. Please note for the purpose of this chapter we refer to the tethered DNA (of known identity) on the microarray slide as the **probe**, and the fluorescently labelled cDNA (synthesized from unknown mRNA messages) as the hybridization **targets**.

The steps involved in microarray analysis are summarized in *Figure 1* (see also *Plate 28*). In brief it involves six steps:

1. Fabrication of cDNA microarrays (choosing cDNA clones, probe production, and printing).
2. Fluorescent target production (RNA extraction and labelling).
3. Hybridization.
4. Image acquisition.
5. Image processing (normalization and data analysis).
6. Statistical analysis and data mining.

The protocols are meant to be a general guide and other kits, reagents, and protocols may be substituted.

2 Fabrication of cDNA microarrays

The first step is to generate a catalogue of genes whose expression levels are being investigated and the cDNA bacterial clones obtained. In general publicly available cDNA clones have been designated a unique identification number by the 'Integrated Molecular Analysis of Genomes and their Expression' (IMAGE) consortium (http://image.llnl.gov/). This consortium was initiated in 1993 by four academic groups on a collaborative basis. IMAGE clones and associated products may be obtained from any of five authorized distributors including,

in the USA:

(a) American Type Culture Collection (http://www.atcc.org/)
(b) Incyte Genomics (http://reagents.incyte.com/)
(c) Research Genetics (http://www.resgen.com/)

In Europe, IMAGE distributors are:

(d) UK HGMP, Hinxton, England (http://www.hgmp.mrc.ac.uk/)
(e) Resource Center of the German Human Genome Project, Berlin Germany (http://www.rzpd.de/)

Prefabricated high-density microarrays can be purchased from a number of sources such as TeleChem (http://www.arrayit.com) and companies such as Incyte (http://www.incyte.com/) offer microarray hybridization and analysis services to investigators who provide the RNA.

2.1 Culturing cDNA bacterial clones

The bacterial clones are obtained and expanded in culture media, the plasmid DNA is extracted, the inserts are amplified by PCR, and purified. All these

224

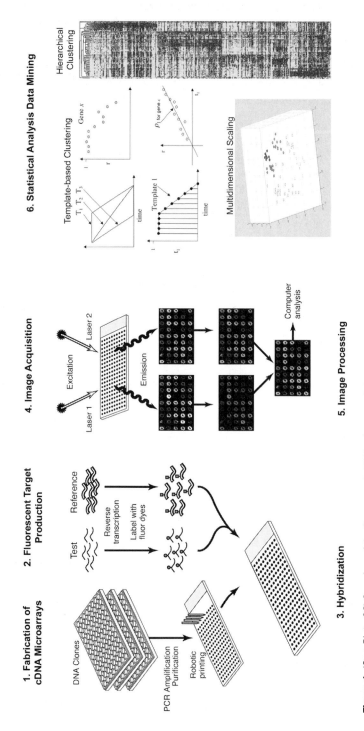

Figure 1 (*See Plate 28*) Overview of the fluorescent cDNA microarray hybridization procedure.

procedures are carried out in the same 96- or 384-well formats as the originating bacterial clones which allows automation. Currently upwards of 20 000 elements can be spotted on a standard microscope slide, which can include all known genes as well as several thousand unknown ESTs, with the major constraint being the surface area of the glass slide.

Protocol 1

Culturing clones for arrays

Equipment and reagents

- 96-well round-bottom plates (Corning Inc., 3799)
- ThinSeal plate sealers (Excel Scientific, STR-THIN-PLT)
- 96-well culture blocks (Edge BioSystems, 4050066)
- Airpore tape sheets (Qiagen, 19571)
- 96-pin inoculation stamp
- LB broth (BSI, 359-000)
- Superbroth (BSI, 371-000)
- Carbenicillin
- Ampicillin
- 100% ethanol

A. Pre-growth of clones to ensure maximum plasmid production[a]

1 In sterile 96-well round-bottom plates, add 100 μl LB broth per well with 100 μg/ml carbenicillin.

2 Thaw frozen 96-well library plates containing source bacterial cultures and spin briefly, 2 min at 200 g, to remove condensation and droplets from the sealer.

3 Sterilize the 96-pin inoculation stamp between samples using 100% ethanol and flame the pins using appropriate safety precautions.

4 After briefly allowing the inoculation block to cool, dip the pins in the library plate, then inoculate the equivalent LB plate ensuring correct orientation. Sterilize the inoculation pins as in step 3.

5 Reseal the library plates with plate sealers (ThinSeal). Store the library plates at −70 °C.

6 Incubate growth plates in humidified oven overnight at 37 °C.

B. Inoculating deep well culture blocks

1 Add 1 ml of Superbroth, containing 100 μg/ml carbenicillin, to each well of the 96-well culture blocks using an 8-channel pipettor.

2 Using the 96-pin inoculation stamp, as above, inoculate the 96-well culture blocks.

3 Cover (Airpore sheets) and place blocks in the 37 °C shaker incubator (200 r.p.m.) for 24 h.

[a] Use extreme care to avoid cross-contamination. Cross-contamination will be evident when PCR products are gel electrophoresed and present multiple bands. Contaminated clones will require re-streaking and sequencing.

Protocol 2

Isolation of plasmid DNA

Reagents

- 96-well alkaline lysis miniprep kit (Edge BioSystems, 91528) (store all buffers at 4°C)
- 1 M Tris–HCl pH 8.0 (Invitrogen, 5568UA)
- 0.5 M EDTA pH 8.0 (Research Genetics, 750009)
- 100% ethanol

Method

1 Isolate plasmid DNA from the cultures using the miniprep kit according to the manufacturer's protocol.

2 Resuspend the plasmid DNA in 200 μl T low E (10 mM Tris–HCl, 0.1 mM EDTA).

3 Store DNA at −20°C and use as template for PCR amplification (see *Protocol 3*).

The isolated DNA is used as a template for PCR amplification with vector primers (AEK M13) using a 96-well format, typically 12 plates at a time.

Protocol 3

PCR amplification of clones

Equipment and reagents

- Cycleplate thin-wall PCR plate (Robbins Scientific, 1038-00-0)
- Cycleseal PCR plate sealer (Robbins Scientific, 1038-00-0)
- MJ Research (DNA Engine Tetrad) PTC-225 Peltier thermal cyclers
- 10 × PCR buffer (Perkin Elmer, N808-0189) (4°C)
- dATP, dCTP, dGTP, dTTP, 100 mM each (Pharmacia, 27-2035-02): store frozen, −20°C
- Ampli Taq polymerase (Perkin Elmer, N808-4015) (−20°C)
- DEPC H_2O (Research Genetics, 750024)
- AEK M13 forward (F) and reverse (R) primers, a custom oligo (Midland Certified) (−20°C)
- AEK M13F 5′ GTTGTAAAACGACGGCCAGTG 3′: stock concentration 1 mM
- AEK M13R 5′ CACACAGGAAACAGCTATG 3′: stock concentration 1 mM

Method

1 A PCR reaction mix is made by combining the following on ice:[a]

Reagent	[Stock]	[Final]	Volume per 1000 reactions
PCR buffer	10 ×	1 ×	10 ml
dATP	100 mM	0.2 mM	0.2 ml
dTTP	100 mM	0.2 mM	0.2 ml
dGTP	100 mM	0.2 mM	0.2 ml

Reagent	[Stock]	[Final]	Volume per 1000 reactions
dCTP	100 mM	0.2 mM	0.2 ml
AEK M13F	1000 μM	0.5 μM	0.05 ml
AEK M13R	1000 μM	0.5 μM	0.05 ml
Ampli Taq Pol	5 U/μl	0.05 U/μl	1 ml
DEPC H$_2$O			87.1 ml

2   Using a multichannel pipette, transfer 99 μl of the master mix to each well of PCR plate (Cycleplate).

3   Using a multichannel pipette, transfer 1 μl of appropriate template DNA in each well taking care to keep the plate orientation and order.[b]

4   Cover the plates with sealers (Cycleseal) and place in thermocycling device.

5   Amplify the templates using the following cycle conditions:

Step	Temperature	Time
1.	96 °C	30 sec
2.	94 °C	30 sec
3.	55 °C	30 sec
4.	72 °C	150 sec
5.	Repeat steps 2–4, 24 times	
6.	72 °C	5 min

[a] It is recommended to make a slight excess of PCR reaction mix than what is actually required.

[b] Take care to remove air bubbles and ensure proper mixing of reaction mix.

Protocol 4

Quantification of PCR product

Equipment and reagents

- FluoReporter blue fluorometric dsDNA kit (Molecular Probes, F-2962) (4 °C)
- Microfluor 2 white 96-well U-bottom plates (Dynex, 7105)
- Lambda *Hind*III fragments (Invitrogen, 15612-013) (−20 °C)
- Perkin Elmer luminescence spectrometer LS50B

Method

1   2 μl of each PCR product is analysed by electrophoresis on a 2% TAE agarose gel containing 0.5 μg/ml ethidium bromide. We obtain a digital image of the gel under UV illumination and analyse the electrophoresis products to ensure that a single band of distinct size is produced for each sample. The intensity of the band gives an estimate of the relative amount of product.

Protocol 4 continued

2 Quantification of PCR products can be accomplished using fluorometric quantitation.[a]

3 Expected yield is ~100 µg/ml.

[a] We use the FluoReporter blue fluorometric dsDNA kit, Dynex Microfluor plates, lambda *Hind*III fragments for standards, and a Perkin Elmer Luminescence Spectrometer LS50B.

Protocol 5

PCR product purification

Equipment and reagents

- 96-well V-bottom plates (Corning Inc., 3894)
- Cyclefoil plate sealers (Robbins Scientific, 1044-39-3)
- Super T21 Centrifuge (Sorvall)
- 1575 ImmunoWash (Bio-Rad)

- Quart size heat sealable bags (Kapak, 404)
- Electric sealer
- 100% ethanol, 70% ethanol
- 3 M sodium acetate pH 6.0
- 3 × SSC

Method

1 Prepare an ethanol/acetate precipitation mix (150 mM sodium acetate pH 6 in ethanol). Add 200 µl of the precipitation mix to each well of a V-bottom 96-well plate.

2 Using multichannel pipettor, transfer the remaining (approx. 97 µl) PCR products to their corresponding wells containing the precipitation mix.[a]

3 Place plates at −80 °C for 1 h, or overnight at −20 °C, to precipitate the DNA.

4 Allow plates to thaw (to reduce brittleness and melt any ice) and spin in a high speed swinging holder centrifuge (Sorvall Super T21). We typically spin stacks of three plates at 1600 g for 1 h.[b] After centrifugation, remove supernatant from plates and dispense 70% ethanol wash, 150 µl per well, using a plate processing station such as the Bio-Rad 1575 ImmunoWash.

5 Centrifuge the plates as in step 4 at 1600 g for 1 h, and remove the supernatant. In a dust-free area, allow plates to dry overnight, without lids and covered with clean paper towels.

6 Resuspend the PCR products in 40 µl of 3 × SSC. Seal plates with foil sealer, making sure that all wells are tightly sealed. Place the plates in an airtight heat sealed bag with a moistened paper towel and place in a 65 °C oven for 2 h.[c]

7 Remove the cooled plates and store at −20 °C.

[a] Ensure proper mixing of the PCR solution with the precipitation mix. Failure to do so will decrease DNA yield concentration.

[b] Use rubber pads between the stacked plates to prevent cracking and breakage.

[c] Heat and cool plates slowly to prevent condensation on the sealer and upper rim of the well.

Protocol 6

Poly-L-lysine pre-treatment of glass slides

Treatment of slides with a coat of poly-L-lysine allows the target DNA to adhere to the surface and minimize loss during hybridization.[a]

Equipment and reagents

- Gold Seal slides (these slides have consistently low intrinsic fluorescence) (Becton Dickinson, 3011)

- 50-slide stainless steel slide racks and glass tanks (Wheaton, 900400)

- Sodium hydroxide (pellets)

- 100% ethanol (the source alcohol should be examined in a fluorometer to ensure that it has very low levels of contaminating fluorescent organic compounds)

- 25-slide plastic slide racks and plastic tanks with lids (Shandon Lipshaw, 195 and 196)

- 0.1% (w/v) poly-L-lysine (Sigma, P8920)

- Tissue culture PBS: sodium chloride 8 g/litre, potassium chloride 0.2 g/litre, sodium phosphate dibasic anhydrous 1.44 g/litre, potassium phosphate monobasic 0.24 g/litre (sterilized and filtered)

Method

1 Place new Gold Seal microscope slides into a stainless steel 50-slide rack.

2 Prepare cleaning solution in large glass beaker (500 ml required per 50-slide glass tank):
   - 400 ml ddH$_2$O
   - 100 g NaOH
   - 600 ml of 95% ethanol

   Dissolve NaOH in water, then add ethanol. Stir until solution is clear. If the solution does not clear, add H$_2$O until it does.

3 Dispense cleaning solution into 50-slide glass tanks. Submerge the rack in the cleaning solution and shake for 2 h on orbital shaker.

4 Remove slides and rinse with fresh ddH$_2$O for 2–5 min. Repeat wash four times, each time using fresh ddH$_2$O.[b]

5 Move clean slides to 25-slide plastic racks.

6 Prepare poly-L-lysine solution as follows (for two boxes of 25 slides each):
   - 35 ml poly-L-lysine (0.1%, w/v)
   - 35 ml tissue culture PBS
   - 280 ml ddH$_2$O

7 Dispense poly-L-lysine solution to plastic 25-slide containers. Submerge rack in poly-L-lysine solution, cover with lid, and shake for 1 h.

8 Rinse once in ddH$_2$O for 1 min.

9 Centrifuge rack in a low speed swinging holder centrifuge to remove free liquid.

10 Immediately transfer to a clean slide box.

11 Allow slides to age for two weeks before printing.[c]

[a] It is important to wear powder-free gloves at all times and avoid contact with detergents or other compounds that may cause background fluorescence.

[b] It is important to remove all traces of cleaning solution. Failure to do so will hinder poly-L-lysine coating reaction and will adversely affect microarray results (low intensity spots, high background).

[c] Aged slides will be very hydrophobic (water drops leave no trail when they move across the surface).

2.2 Microarray slide printing

The next stage is printing of DNA probes on the coated glass slides. The printing process refers to the robot-driven sequential transfer of individual purified PCR amplified fragments from a 96-well microtitre tray to exact predefined locations on glass slides. Several arrayers are available from commercial companies:

(a) Affymetrix 417 Arrayer (http://www.affymetrix.com/products/spotted.html)

(b) Cartesian Technologies (http://www.cartesiantech.com/)

(c) Beecher Instruments (http://www.beecherinstruments.com/)

(d) Genomic Solutions (http://www.genomicsolutions.com/)

(e) BioRobotics (http://www.biorobotics.co.uk/)

It is also possible to build your own arrayer (http://cmgm.stanford.edu/pbrown/), and arrayers have been built in several academic settings (12).

After printing is complete, etch identifying marks along the top of each slide (print number and slide number) with a diamond scriber and place slides in plastic slide box (use simple plastic slide boxes with no paper or cork to shed particles). One slide should be marked for use as a template to line up the cover-slip for microarray hybridizations, as the printed spots will not be visible after the slide blocking procedure.

The slides are aged one week, blocked, and the printed DNA denatured. The blocking reduces non-specific binding of strongly negatively charged target on microarray slides, the positively charged amine groups on poly-L-lysine coated slides are passivated by reaction with succinic anhydride. We routinely process 48 slides at a time.

Protocol 7

Blocking slides after printing with succinic anhydride

Equipment and reagents[a]

- Stratagene UV Stratalinker 2400
- 30-slide stainless steel slide rack (Wheaton, 900234)
- 30-slide glass submersion tanks (Wheaton, 900303)
- 500 ml glass beakers and stir bars
- Large glass dish (14 inch casserole)
- Large round Pyrex dishes (8 inch diameter)

- Plastic slide box
- 1-Methyl-2-pyrrolidinone (Aldrich, 32 863-4)
- Succinic anhydride (Aldrich, 23 969-0)
- 1 M sodium borate pH 8.0 (adjust pH of boric acid with sodium hydroxide)
- 100% ethanol

Method

1. Age slides for one week at room temperature after printing.[b]

2. Place slides in glass casserole dish and cover with plastic wrap. UV crosslink printed cDNA with a dose of 450 mJ UV energy (Strategene Stratalinker).

3. Place slides in stainless steel 30-slide racks, and place racks in clean glass tanks. Prepare passivation reaction (for one tank) in dedicated, dry, 500 ml beaker:
   - Succinic anhydride   6 g
   - 1-Methyl-2-pyrrolidinone   325 ml
   - 1 M sodium borate pH 6.0   25 ml

   When succinic anhydride has completely dissolved, add 25 ml of 1 M sodium borate buffer while mixing and quickly pour on to slides.[c]

4. Shake the slides for 20–30 min on orbital shaker—some precipitation will occur. While blocking, boil ddH$_2$O in a clean Pyrex dish using a hot plate, so that it will be ready after the reaction.[d] Remove slide holder from the passivation reaction and dunk immediately in boiling ddH$_2$O to denature DNA. Turn off heat source and let stand for 2 min in the nearly boiling ddH$_2$O bath. Remove slide holder and dunk in fresh glass tank with 100% ethanol to dehydrate slides.

5. After 3–5 min in ethanol, remove the slides and centrifuge dry in a low speed swinging holder centrifuge.

6. Place dry slides in a clean slide box.

[a] The reaction solution must be prepared in completely dry containers. All glassware, stir bars, and graduated cylinders should be dedicated to slide blocking and should not be cleaned with detergents, which can adversely affect this reaction.

[b] A number of groups have found that rapid or slow hydration of the DNA on the slide after printing, followed by a quick drying step improves DNA distribution or signal strength. This has not been observed for materials prepared by our procedure, so it is routinely omitted.

[c] When water is added, the anhydride will begin to rapidly decompose, so add the mix to the slides very quickly. It is helpful to dispense the sodium borate buffer from a pre-aliquoted 50 ml conical tube.

[d] Cover dish with aluminium foil to reduce evaporation and simultaneously boil ddH$_2$O in microwave to replenish evaporative loss.

3 Target production

The purity and integrity of the total RNA samples are vital factors for successful cDNA microarray analysis. Gloves must be worn and no touch techniques followed at all times. We double purify the RNA to ensure minimal DNA, protein, and carbohydrate contamination. For cell lines, RNA first extracted using a Qiagen RNeasy kit followed by a further round of purification using Trizol yields excellent results (two rounds of Trizol extraction is also accepted and is recommended for tissues). The amount of total RNA in each channel required for a microarray experiment varies from 50–200 µg, with the precise amount varying with the size of the array and fluorescent nucleotide used.

Considerable thought should be given to what reference cell line(s) or tissue to use for your microarray experiments. The reference should be abundant and offer at least a minimal intensity for all genes printed on your array (if a gene has no intensity in the reference channel, then ratios and other statistical calculations cannot be computed as the denominator cannot be zero).

Protocol 8

RNA extraction[a]

Equipment and reagents

- Virsonic 100 with micro probe: conical titanium probe, 1.8 mm diameter tip (Virtis, 346411)
- RNeasy Midi Kit (Qiagen, 75142)
- Trizol (Invitrogen, 15596-018) (4 °C)
- 2-Mercaptoethanol (ICN Biomedicals, 806445)

- Microcon-30 (Millipore, 142410)
- DEPC H_2O (Research Genetics, 750024)
- Chloroform
- Isopropanol
- 50 mM sodium hydroxide
- 70% ethanol

Method

1 For the RNeasy kit follow the manufacturer's guidelines. At the final elution stage, elute with two successive aliquots of 150 µl RNase-free water.[b]

2 If starting with tissue, for each 100 mg of frozen tissue add to 5 ml of Trizol, and dissociate by homogenization with a rotating blade tissue homogenizer. Follow the manufacturer's guidelines for RNA extraction. Dissolve RNA pellet in 100 µl of DEPC treated water.

3 Determine the concentration of your RNA in 50 mM NaOH.

4 At this time it may be convenient to aliquot out appropriate quantities (55–220 µg) of your RNA before beginning the second round of purification.[c]

5 Extract the eluted RNA a second time by adding 1 ml of Trizol per maximum 0.1 ml of RNA, vortexing, and follow manufacturer guidelines for RNA extraction.

6 You may leave the precipitated RNA in isopropanol at $-80\,^{\circ}$C for later use.

7 If the RNA is to be used immediately, wash the RNA pellet twice with 70% ethanol, remove the ethanol, and dry the pellet (air dry or Speed Vac).[d]

8 Resuspend the dried RNA pellet (50–200 μg) in 400 μl RNase-free H_2O. Take 1 μl for final RNA concentration measurement, and make sure the total amount of RNA for labelling with Cy3 is 80–100 μg, and 150–200 μg for Cy5.

9 Transfer RNA to Microcon-30 and centrifuge at 14 000 g for 7–12 min to concentrate RNA. Concentrate RNA to less than 14 μl.[e]

10 Elute RNA, and bring to a final volume of 14 μl with DEPC H_2O.

[a] High quality RNA is crucial to the success of a microarray experiment. It is also possible to use 2–4 μg of poly(A)-purified mRNA in the target synthesis reaction. Smaller quantities of RNA may be used in conjunction with RNA amplification techniques, which are currently under development.

[b] Cell lines should be harvested under consistent conditions and lysed rapidly. It is recommended that you sonicate the very viscous lysate with several 5 sec bursts to disrupt the genomic DNA before applying to the RNeasy column. We use the micro probe (conical titanium probe 1.8 mm diameter tip) at a setting 5, dissipated power approx. 5–10 watts. When extracting RNA from tissue add the frozen sample directly to the Trizol without thawing with immediate homogenization.

[c] Factor in an approx. 10% loss of RNA from each Trizol purification round.

[d] Over-drying may make resuspension of RNA difficult and may adversely affect results.

[e] Do not concentrate to dryness, as the sample may be lost or difficult to recover. Proper volume is when the filter is partially dry upon visual inspection.

Following extraction from the samples the RNA is reverse transcribed using anchored poly(dT) primers in the presence of fluor-derivatized nucleotides. Currently the factors of labelling efficiency, fluorescent yield, spectral separation, and non-specific binding make the Cy3/Cy5 pair the most useful for our detection system. Although there are a number of conjugated fluorophores available (dCTP, dUTP, amino-allyl dUTP RT coupled to monofunctional dyes), we have found that Amersham Pharmacia dUTP-conjugated Cy3/Cy5 yields consistent results. This procedure provides the tagged representations of the mRNA pools in the form of single-stranded cDNA that will be hybridized to the immobilized cDNA PCR products on the microarry. Other labelling systems are being tested.

Protocol 9

Direct labelling of cDNA with fluorescent dyes

Reagents

- 100 mM dNTPs (Amersham Pharmacia, 27-2035-02)

- 10 × low T dNTPs nucleotide mix: 5 mM dGTP, 5 mM dATP, 5 mM dCTP, 2 mM dTTP (store at −20 °C)

- FluoroLink Cy3-dUTP 1 mM (Amersham Pharmacia, PA53022), photosensitive (−20 °C)[a]

- FluoroLink Cy5-dUTP 1 mM (Amersham Pharmacia, PA55022), photosensitive (−20 °C)[a]

- SuperScript II reverse transcriptase enzyme, 5 × first strand buffer, 0.1 M DTT (Invitrogen, 18064-014) (−20 °C)

- Anchored oligo(dT) primer, 1 µg/µl: d-20T-d(AGC) (−20 °C)

- RNase inhibitor RNasin (Promega, N211A) (−20 °C)

- DEPC H_2O (Research Genetics, 750024)

Method

1 Pre-anneal 14 µl RNA with 3 µl anchored oligo(dT) primer, 1 µg/µl. Incubate in thermocycler at 70 °C for 5 min and cool to 42 °C.

2 Mix the RT labelling mix:

   - Cy3 *or* Cy5-dUTP (1 mM)   4 µl
   - 5 × first strand buffer   8 µl
   - 10 × low T dNTP mix   4 µl
   - 0.1 M DTT   4 µl
   - RNase inhibitor RNasin   1 µl
   - SSII reverse transcriptase   2 µl

3 Add RT labelling mix to the pre-annealed RNA.

4 Incubate at 42 °C for 30–60 min.

5 Add a further 2 µl SSII RT enzyme, incubate at 42 °C for another 30–60 min, and cool to room temperature.

[a] Cye dyes are also available from NEN Life Sciences.

Protocol 10

Target purification

The labelled target reaction must be purified to remove unincorporated nucleotides.

Reagents

- 0.5 M EDTA pH 8.0 (Research Genetics, 750009)

- 1 M NaOH

- 1 M Tris–HCl pH 7.5 (Quality Biological, 351-006-100)

Method

1   To stop the labelling reaction, add 5 μl of 0.5 M EDTA pH 8.0 and mix well.

2   To hydrolyse the RNA, add 10 μl of 1 M NaOH and mix well. Incubate at 65 °C for 20–30 min, then cool to room temperature.

3   Add 25 μl of 1 M Tris–HCl pH 7.5 to neutralize the NaOH.

4   Purify each labelled colour individually for the first purification. In Microcon-30 spin columns, add labelled target and bring up to a total volume of 400 μl with DEPC H_2O. Spin column at 14000 g for about 8–9 min to a volume of approx. 50 μl.[a]

5   Recover each target. For the second purification, pool the Cy3 and Cy5 labelled targets for an experiment in a new Microcon-30 column and bring up to a total volume of 400 μl H_2O.

6   Concentrate the combined targets to 25 μl final volume for hybridization.

[a] The flow-through may be saved at this step for HPLC recovery of unincorporated fluorophores.

4 Hybridization

Hybridization volumes may vary depending on array size. *Protocol 11* is based on a 20 × 40 mm array. Adjust volumes proportionally and use appropriate sized coverslips for smaller/larger arrays.

Protocol 11

Microarray hybridization

Equipment and reagents

- Slide hybridization chamber (TeleChem, AHC-1)
- 10 mg/ml poly(dA) (Sigma, P9403) (−20 °C)
- 4 mg/ml yeast tRNA (Invitrogen, 15401-011) (−20 °C)
- Human[a] Cot-1 DNA (concentrated to 10 mg/ml) (Invitrogen, 15279-011) (−20 °C)
- 50 × Denhardts (Research Genetics, 750018) (−20 °C)
- 10% SDS (Research Genetics, 750008)
- 20 × SSC

Method

1   Make hybridization mixture containing competitor DNA (to reduce non-specific binding and background):
   - Pooled Cy5/Cy3 labelled targets   25 μl
   - 10 mg/ml poly(dA)   1.5 μl
   - 4 mg/ml yeast tRNA   1.5 μl
   - 10 mg/ml human[a] Cot-1 DNA   1.5 μl
   - 50 × Denhardts   1.5 μl
   - 20 × SSC   5 μl

Protocol 11 continued

2 Denature at 98 °C for 2 min and cool on wet ice for 10 sec.

3 Add 0.8 μl of 10% SDS.

4 Pipette hybridization targets up and down several times until well mixed and place mixture on microarray under a 24 × 50 mm glass coverslip.[b]

5 Place microarray slide in hybridization chamber with 15–20 μl of 3 × SSC to maintain humidity within the chamber.

6 Incubate microarray hybridization chamber in 65 °C water-bath for 12–18 h.

[a] Appropriate competitor DNA should be used for microarrays of clones from other organisms, such as mouse.

[b] For the best results, apply pooled targets to the centre of the coverslip and then, using the template slide as a guide (see Section 2.2), place inverted microarray slide from above. Avoid air bubbles.

After hybridization, the hybridization solution and any unbound target must be removed from the surface of the slide to reduce background.

Protocol 12

Post-hybridization slide washes

Equipment and reagents

- Glass Coplin staining jars (Wheaton, 900470)
- 25-slide plastic slide racks (Shandon Lipshaw, 195)
- Wash solution 1: 0.1% SDS, 0.5 × SSC in ddH$_2$O (filtered)
- Wash solution 2: 0.01% SDS, 0.5 × SSC in ddH$_2$O (filtered)
- Wash solution 3: 0.06 × SSC in ddH$_2$O (filtered)

Method

1 Remove microarray hybridization chamber from water-bath.[a]

2 Dispense wash solutions into Coplin staining jars. Open hybridization chamber and immediately submerge entire slide in wash 1 until coverslip slips off. Once coverslip comes off, agitate gently for 2 min.[b]

3 Transfer slide to wash 2 and agitate gently for 2 min.

4 Transfer slide to wash 3 and agitate gently for 2 min.

5 Place the slide in a plastic 25-slide rack and spin in a centrifuge equipped with a swinging carrier (horizontal) which can hold the slide holder. Spin immediately (900 g for 2 min at room temperature).

6 Scan slide as soon as possible.

[a] Use paper towels and vacuum suction to completely dry outer surface of chamber.

[b] Take care that the coverslip does not scratch the microarray surface.

5 Image acquisition

Target fluorescence intensities at the immobilized probes can be measured using a variety of commercially available scanners (13). The following is a brief list of commercially available scanners and contact information.

(a) Affymetrix: 418 Array Scanner—scanning laser digital imaging epifluorescence microscope. 532 nm (35 mW) and 635 nm (35 mW) lasers with 3 min scan time per slide. (http://www.affymetrix.com/)

(b) Agilent: under development. (http://www.agilent.com/)

(c) Axon: GenePix 4000—532 nm (20 mW) and 635 nm (15 mW) lasers and 10 μm pixel resolution. 5 min scan time per slide. (http://www.axon.com/)

(d) Beecher Instruments: scanner—laser confocal, two simultaneous PMT channels. Three lasers: 488 nm @ 75 mw, 532 nm @ 100 mw, 633 nm @ 35 mw. 10–100 μm pixel resolution. (http://www.beecherinstruments.com/)

(e) Genomic Solutions: GeneTAC LS IV and GeneTAC 2000—CCD camera with high energy xenon light source. Can scan up to four fluors per slide. (http://www.genomicsolutions.com/)

(f) GSI Lumonics: ScanArray LITE, ScanArray 4000, ScanArray 5000—scanning confocal laser GHeNe 543 nm (Cy3) and RHeNe 632 nm (Cy5). (http://www.genscan.com/)

(g) Molecular Dynamics: array scanner—confocal optics, nine element lens. HeNe and NdYag lasers. Scanning time per slide: 5 min single colour, 11 min two colour. (http://www.moleculardynamics.com/)

(h) Packard Instruments: BioChip Imager—epifluorescence confocal scanning laser system. 543 nm (Cy3) and 633 nm (Cy5) HeNe lasers with 50 μm, 20 μm, or 10 μm pixel resolution. (http://packardinst.com/)

6 Image analysis and normalization

The two image files generated by the scanner are analysed using software tools (Array Suite) developed by Chen *et al.* for the ScanAlytics IPLab image processing package (14). These software tools can be used with any image file format to extract raw target intensity information as well as compute background-corrected intensities and expression ratios, confidence intervals, and allow for data integration of all clone information. As each probe is roboticly printed to a predefined position, the scanned images are overlaid with a grid that divides the images into segments, each containing a target spot. All clone information, including gene name, clone identification number, chromosome and radiation hybrid mapped location, and source microplate position, is attached to each segment by this process. Each of the images are assigned a pseudo-colour (e.g. Cy5 = red and Cy3 = green). The probe spot is identified within each segment and the target fluorescence intensity is calculated for each colour by averaging the intensities of every pixel inside the detected spot region. The local back-ground intensity around each spot in each colour is also measured within each segment. For every spot in each colour channel, the final target intensity values are derived by subtracting the local background intensity from the average fluorescence intensity.

Next, a normalization constant is determined to compensate for differential efficiencies of labelling and detection of Cy3 and Cy5. The process involves calculating the average intensity, in both colour channels, for a set of internal controls which are housekeeping genes. These genes are pre-selected and have been verified on numerous hybridizations as being stable for most experiments (red/green ratio = 1.0). The normalization constant is then derived and used to calculate a calibrated red/green ratio for each cDNA spot within the image. In addition the ratio variance of the control genes is used to calculate 99% confidence intervals in which the ratios are statistically no different from 1. The output of the analysis is in the form of a pseudo-coloured image of the entire array. Individual spots can be highlighted using the mouse cursor and informa-tion including gene name, clone identity, intensity values, intensity ratios, normalization constant, and user-defined confidence intervals can be obtained. A spreadsheet of expression ratio data for each spot is generated. For more information about this process and to download free software and tools, see our web site (http://www.nhgri.nih.gov/DIR/LCG/15K/HTML/).

7 Sensitivity and specificity

It is estimated that the sensitivity of this method allows the detection of mRNA species comprising 1:10 000 of the mass of poly(A^+). Comparisons between the microarray experiments with Northern hybridizations have confirmed this tech-nique to be reliable (15–18). Our experience to date has indicated the high consistency of microarray data for determining ratio changes.

8 Data mining and statistical analysis

All data from each experiment can be downloaded into a relational database such as FileMaker Pro (Claris) and further parsed for comparing data across experiments as well as for extracting data from individual array hybridizations. It is obvious that large scale, high-throughput experimental methods require information processing coupled to a variety of analysis tools. Software tools such as ArrayDB (http://www.nhgri.nih.gov/DIR/LCG/15K/HTML/) (19) can also be used to integrate information from many Internet sources, such as NCBI Entrez, UniGene, and KEGG databases, with experimental gene expression data. Hier-archical clustering of biological samples and genes (16, 20, 21) is a commonly applied mathematical strategy to organize gene expression data. Algorithms such as multidimensional scaling are proving to be an informative way to visualize expression profiles (16). More complex data analysis systems (22–25) are currently being devised for complex clustering of data, a description of this is beyond the scope of this review (*Figure 1*; see also *Plate 28*).

9 Summary

The range of applications for the cDNA microarray technology is as broad as the imagination. For example the temporal impact on gene expression by drugs (26), environmental toxins (27), or oncogenes (17) may be elucidated and regulatory networks and co-expression patterns can be deciphered. In addition recent studies in human cancer demonstrated that DNA microarrays have been utilized to develop novel molecular taxonomies of cancer (28) including clustering of cancers according to prognostic groups on the basis of gene expression profiles (29). The list of potential uses of this technique is not limited to cancer research and in the five years since its inception it has become a major tool for the investigation of global gene expression of all aspects of human disease and in biomedical research (16).

References

1. Collins, F. S., Patrinos, A., Jordan, E., Chakravarti, A., Gesteland, R., and Walters, L. (1998). *Science*, **282**, 682.
2. Roest Crollius, H., Jaillon, O., Bernot, A., Dasilva, C., Bouneau, L., Fischer, C., *et al.* (2000). *Nature Genet.*, **25**, 235.
3. Liang, F., Holt, I., Pertea, G., Karamycheva, S., Salzberg, S. L., and Quackenbush, J. (2000). *Nature Genet.*, **25**, 239.
4. Ewing, B. and Green, P. (2000). *Nature Genet.*, **25**, 232.
5. Fields, C., Adams, M. D., White, O., and Venter, J. C. (1994). *Nature Genet.*, **7**, 345.
6. Antequera, F. and Bird, A. (1993). *Proc. Natl. Acad. Sci. USA*, **90**, 11995.
7. Velculescu, V. E., Zhang, L., Vogelstein, B., and Kinzler, K. W. (1995). *Science*, **270**, 484.
8. Lockhart, D. J., Dong, H., Byrne, M. C., Follettie, M. T., Gallo, M. G., Chee, M. S., *et al.* (1996). *Nature Biotechnol.*, **14**, 1675.
9. Drmanac, S. and Drmanac, R. (1994). *Biotechniques*, **17**, 328, 332.
10. Drmanac, S., Stavropoulos, N. A., Labat, I., Vonau, J., Hauser, B., Soares, M. B., *et al.* (1996). *Genomics*, **37**, 29.
11. Schena, M., Shalon, D., Davis, R. W., and Brown, P. O. (1995). *Science*, **270**, 467.
12. Cheung, V. G., Morley, M., Aguilar, F., Massimi, A., Kucherlapati, R., and Childs, G. (1999). *Nature Genet.*, **21**, 15.
13. Bowtell, D. D. (1999). *Nature Genet.*, **21**, 25.
14. Chen, Y., Dougherty, E. R., and Bittner, M. L. (1997). *Biomed. Opt.*, **2**, 364.
15. DeRisi, J., Penland, L., Brown, P. O., Bittner, M. L., Meltzer, P. S., Ray, M., *et al.* (1996). *Nature Genet.*, **14**, 457.
16. Khan, J., Bittner, M. L., Chen, Y., Meltzer, P. S., and Trent, J. M. (1999). *Biochim. Biophys. Acta*, **1423**, M17.
17. Khan, J., Bittner, M. L., Saal, L. H., Teichmann, U., Azorsa, D. O., Gooden, G. C., *et al.* (1999). *Proc. Natl. Acad. Sci. USA*, **96**, 13264.
18. Khan, J., Saal, L. H., Bittner, M. L., Chen, Y., Trent, J. M., and Meltzer, P. S. (1999). *Electrophoresis*, **20**, 223.
19. Ermolaeva, O., Rastogi, M., Pruitt, K. D., Schuler, G. D., Bittner, M. L., Chen, Y., *et al.* (1998). *Nature Genet.*, **20**, 19.

20. Spellman, P. T., Sherlock, G., Zhang, M. Q., Iyer, V. R., Anders, K., Eisen, M. B., *et al.* (1998). *Mol. Biol. Cell*, **9**, 3273.

21. Eisen, M. B., Spellman, P. T., Brown, P. O., and Botstein, D. (1998). *Proc. Natl. Acad. Sci. USA*, **95**, 14863.

22. Toronen, P., Kolehmainen, M., Wong, G., and Castren, E. (1999). *FEBS Lett.*, **451**, 142.

23. Brown, M. P., Grundy, W. N., Lin, D., Cristianini, N., Sugnet, C. W., Furey, T. S., *et al.* (2000). *Proc. Natl. Acad. Sci. USA*, **97**, 262.

24. Gaasterland, T. and Bekiranov, S. (2000). *Nature Genet.*, **24**, 204.

25. Aach, J., Rindone, W., and Church, G. M. (2000). *Genome Res.*, **10**, 431.

26. Marton, M. J., DeRisi, J. L., Bennett, H. A., Iyer, V. R., Meyer, M. R., Roberts, C. J., *et al.* (1998). *Nature Med.*, **4**, 1293.

27. Afshari, C. A., Nuwaysir, E. F., and Barrett, J. C. (1999). *Cancer Res.*, **59**, 4759.

28. Khan, J., Simon, R., Bittner, M., Chen, Y., Leighton, S. B., Pohida, T., *et al.* (1998). *Cancer Res.*, **58**, 5009.

29. Alizadeh, A. A., Eisen, M. B., Davis, R. E., Ma, C., Lossos, I. S., Rosenwald, A., *et al.* (2000). *Nature*, **403**, 503.

List of suppliers

• Companies are often subject to change in their organizational structure and affiliations leading to alterations to a supplier's web site URL. The editors of this book have therefore established a link at an academic web site that will contain supplier URL updates and also any corrections to errors in the text that may take place subsequent to printing.

Supplier URL updates and minor corrections to text in this book can be found by following the links at: http://www.utoronto.ca/cancyto

Adobe Systems Incorporated, 345 Park Avenue, San Jose, CA 95110-2704, USA.
Tel: 800-445-8787
URL: http://www.adobe.com.type

Affymetrix Headquarters, 3380 Central Exwy, Santa Clara, CA 95051, USA.
Tel: 408-731-5503
Fax: 408-481-9442
URL: http://www.affymetrix.com

Aldrich, 1001 West Saint Paul Avenue, Milwaukee, WI 53233, USA.
Tel: 414-273-3850
Fax: 414-273-4979
URL: http://www.aldrich.com

Ambion RNA Diagnostics, 2130 Woodward Street, Austin, TX 78744-1832, USA.
Tel: 877-777-1874
Fax: 512-651-0201
URL: http://www.ambion.com

Amersham Life Sciences Inc., 2636 S. Clearbrook Drive, Arlington Heights, IL 60005, USA.
Tel: 732-457-8000
Fax: 732-457-0557
URL: http://www.apbiotech.com/na

Amersham Pharmacia Biotech, Inc., 800 Centennial Avenue, PO Box 1327, Piscataway, NJ 08855-1327, USA.
Tel: 732-457-8000   Fax: 877-295-8102
URL: http://www.apbiotech.com

Anderman and Co. Ltd., 145 London Road, Kingston-upon-Thames, Surrey KT2 6NH, UK.
Tel: 0181 541 0035
Fax: 0181 541 0623

Applied Spectral Imaging Inc., 2120 Las Palmas Dr Suited, Carlsbad, CA 92009, USA.
URL: http://www.spectral-imaging.com

Arcturus Engineering Inc., 6001 Arcturus Avenue, Oxnard, CA 93033, USA.
Tel: 800-732-5013   Fax: 805-986-1994
URL: http://www.arcturusmfg.com

ATCC, 10801 University Boulevard, Manassas, VA 20110-2209, USA.
Tel: 703-365-2700   Fax: 703-365-2701
URL: http://www.atcc.org

B. Braun Melsungen AG, D-34209 Melsungen, Germany.
Tel: +49 5661 713394
Fax: +49 5661 713699
URL: http://www.bbraun.de

Beckman Coulter (UK) Ltd., Oakley Court, Kingsmead Business Park, London Road, High Wycombe, Buckinghamshire HP11 1JU, UK.
Tel: 01494 441181
Fax: 01494 447558
URL: http://www.beckman.com
Beckman Coulter Inc., 4300 N Harbor Boulevard, PO Box 3100, Fullerton, CA 92834-3100, USA.
Tel: 001 714 871 4848
Fax: 001 714 773 8283
URL: http://www.beckman.com

Becton Dickinson and Co., 21 Between Towns Road, Cowley, Oxford OX4 3LY, UK.
Tel: 01865 748844
Fax: 01865 781627
URL: http://www.bd.com
Becton Dickinson and Co., 1 Becton Drive, Franklin Lakes, NJ 07417-1883, USA.
Tel: 001 201 847 6800
URL: http://www.bd.com

Beecher Instruments, PO Box 8704, Silver Spring, MD, USA.
Tel/Fax: 301-585-6621
URL: http://www.beecherinstruments.com

Biochrom KG, Leonorenstr. 2-6, D-12247 Berlin, Germany.
Tel: +49 30 77 99 06 0
Fax: +49 30 77 10 01 2
URL: http://www.biochrom.de

Bio 101 Inc., c/o Anachem Ltd., Anachem House, 20 Charles Street, Luton, Bedfordshire LU2 0EB, UK.
Tel: 01582 456666
Fax: 01582 391768
URL: http://www.anachem.co.uk
Bio 101 Inc., PO Box 2284, La Jolla, CA 92038-2284, USA.
Tel: 001 760 598 7299
Fax: 001 760 598 0116
URL: http://www.bio101.com

Bio-Rad Laboratories Ltd., Bio-Rad House, Maylands Avenue, Hemel Hempstead, Hertfordshire HP2 7TD, UK.
Tel: 0181 328 2000   Fax: 0181 328 2550
URL: http://www.bio-rad.com
Bio-Rad Laboratories Ltd., Division Headquarters, 1000 Alfred Noble Drive, Hercules, CA 94547, USA.
Tel: 001 510 724 7000
Fax: 001 510 741 5817
URL: http://www.bio-rad.com

BioRobotics, 185 New Boston Street, Woburn, MA 01801, USA.
Tel: 877-BIOROBO   Fax: 781-376-9792
URL: http://info@biorobotics.com

BioSource International, Inc., 820 Flynn Road, Camarillo, CA 93012, USA.
Tel: 805-987-0086   Fax: 805-987-3385
URL: http://www.biosource.com
BioSource International Europe S.A., Rue de l'Industrie 8, B-1400 Nivelles, Belgium.
Tel: +32 67 88 99 00   Fax: +32 67 88 99 00
URL: http://www.biosource.com
Biosource International (Keystone), 1170B Chess Drive, Foster City, CA 94404, USA.
Tel: 800-788-4362   Fax: 800-786-4362
URL: http://www.keydna.com

Bitplane AG, Badenerstrasse 682, CH-8048 Zürich, Switzerland.
Tel: +41 1 430 11 00
Fax: +41 1 430 11 01
URL: http://www.bitplane.com

Calbiochem/CedarLane, 5516-8th Line, R.R.2, Hornby, Ontario LOP 1EO, Canada.
Tel: 905-878-8891
Fax: 905-878-7800
URL: http://www.cedarlanelabs.com

Cartesian Technologies, Inc., 17851 Sky Park Circle, Suite C, Irvine, CA 92614, USA.
Tel: 800-935-8007
Fax: 949-622-0255
URL: http://www.cartesiantech.com

Cel-Line Associates, Inc., PO Box 648, 33 Gorgo Lane, Newfield, NJ 08344, USA.
Tel: 800-662-0973
Fax: 800-609 0973
URL: http://cel-line.com

Chroma Technology Corporation, 74 Cotton Mill Hill, Unit A9, Brattleboro, VT 05301, USA.
Tel: 802-257-1800
Fax: 802-257-9400
URL: http://www.chroma.com

Corning, One Riverfront Plaza, Corning, NY 14831, USA.
Tel: 607-974-9000
URL: http://www.corning.com

CP Instrument Co. Ltd., PO Box 22, Bishop Stortford, Hertfordshire CM23 3DX, UK.
Tel: 01279 757711
Fax: 01279 755785
URL: http://www.cpinstrument.co.uk

DAKO Corporation, 6392 Via Real, Carpinteria, CA 93013, USA.
Tel: 800-235-5763
Fax: 800-566-3256
URL: http://www.dakousa.com

Dupont (UK) Ltd., Industrial Products Division, Wedgwood Way, Stevenage, Hertfordshire SG1 4QN, UK.
Tel: 01438 734000
Fax: 01438 734382
URL: http://www.dupont.com
Dupont Co. (Biotechnology Systems Division), PO Box 80024, Wilmington, DE 19880-002, USA.
Tel: 001 302 774 1000
Fax: 001 302 774 7321
URL: http://www.dupont.com

Dynex Technologies, 4751 Mustang Circle, St. Paul, MN 55112, USA.
Tel: 763-784-5397
URL: http://www.dynextechnologies.com

Eastman Chemical Co., 100 North Eastman Road, PO Box 511, Kingsport, TN 37662-5075, USA.
Tel: 001 423 229 2000
URL: http://www.eastman.com

Edge BioSystems, 19208 Orbit Drive, Gaithersburg, MD 19208, USA.
Tel: 800-326-2685
Fax: 301-990-0881
URL: http://www.edgebio.com

Eppendorf Vertrieb, Deutschland GmbH, Friedensstrasse 116, 51145 Koln, Germany.
Tel: 0180-325-5911
Fax: 02203-927655
URL: http://www.eppendorf.com

Excel Scientific, PO Box 476, Wrightwood, CA 92397, USA.
Tel: 760-249-6371
Fax: 760-249-6395
URL: http://www.excelscientific.com

Falcon Plastics, 250 West Wylie Avenue, Washington, PA 15301, USA.
Tel: 724-222-2600
Fax: 724-222-4585
URL: http://www.falconplastics.com

Fisher Scientific UK Ltd., Bishop Meadow Road, Loughborough, Leicestershire LE11 5RG, UK.
Tel: 01509 231166
Fax: 01509 231893
URL: http://www.fisher.co.uk
Fisher Scientific, Fisher Research, 2761 Walnut Avenue, Tustin, CA 92780, USA.
Tel: 001 714 669 4600
Fax: 001 714 669 1613
URL: http://www.fishersci.com
Fisher Scientific Ltd., 112 Colonade Road, Nepean, Ontario K2E 7L6, Canada.
Tel: 877-627-7225
URL: http://www.fisherscientific.com

Fluka, PO Box 2060, Milwaukee, WI 53201, USA.
Tel: 001 414 273 5013
Fax: 001 414 2734979
URL: http://www.fluka.com
Fluka Chemical Co. Ltd., PO Box 260, CH-9471, Buchs, Switzerland.
Tel: 0041 81 745 2828
Fax: 0041 81 756 5449
URL: http://www.fluka.com
Fluka Chemicals Ltd., The Old Brickyard, New Road, Gillingham, Dorset SP8 4JL, UK.
Tel: 44 1747 822211   Fax: 44 1747 823779
URL: http://www.fluka.com

FMC BioProducts, Crockett's Point, Box 308, Rockland, ME 04841, USA.
Tel: 207-594-3200   Fax: 207-594-3308
URL: http://www.fmc.com

Genomic Solutions, Inc., 4355 Varsity Drive, Suite E, Ann Arbor, MI 48108, USA.
Tel: 734-975-4800   Fax: 734-975-4808
URL: http://www.genomicsolutions.com

Harlan Sera-Lab Ltd., Hillcrest, Dodgeford Lane, Belton, Loughborough, Leicester LE12 9TE, UK.
Tel: 01530 222123   Fax: 01530 224970

HGMP Resouce Center, Hinxton, Cambridge CB10 1SB, UK.
Tel: +44 (0)1223 494500
Fax: +44 (0)1223 494512
URL: http://www.hgmp.mrc.ac.uk/

Hybaid Ltd., Action Court, Ashford Road, Ashford, Middlesex TW15 1XB, UK.
Tel: 01784 425000
Fax: 01784 248085
URL: http://www.hybaid.com
Hybaid US, 8 East Forge Parkway, Franklin, MA 02038, USA.
Tel: 001 508 541 6918
Fax: 001 508 541 3041
URL: http://www.hybaid.com

HyClone Laboratories, 1725 South HyClone Road, Logan, UT 84321, USA.
Tel: 001 435 753 4584
Fax: 001 435 753 4589
URL: http://www.hyclone.com

ICN Biomedicals, 3300 Hyland Avenue, Costa Mesa, CA 92626, USA.
Tel: 714-545-0100 X3230
Fax: 714 641 7215
URL: http://www.icnbiomed.com
Biochemicals Division, 1263 Chillicothe Road, Aurora, Ohio 44202-8064, USA.
Tel: 330-562-1500
URL: http://www.icnbiomed.com

Incyte Genomics
URL: http://reagents.incyte.com

Invitrogen Canada, 2270 Industrial Street, Burlington, Ontario L7P 1A1, Canada.
Tel: 800-263-6236
Fax: 800-387-1007
URL: http://www.invitrogen.com

Invitrogen Corp., 1600 Faraday Avenue, Carlsbad, CA 92008, USA.
Tel: 001 760 603 7200
Fax: 001 760 603 7201
URL: http://www.invitrogen.com
Invitrogen, 3 Fountain Drive, Inchinnan Business Park, Paisley PA4 9RF, UK.
Tel: 0141 814 6100
Fax: 0141 814 6287
URL: http://www.invitrogen.com
Invitrogen Inc., 1600 Faraday Ave, Carlsbad, CA 92008, USA.
Tel: 001 760 603-7200
Fax: 001 760 602-6500
URL: http://www.invitrogen.com
Invitrogen BV, PO Box 2312, 9704 CH Groningen, The Netherlands.
Tel: 00800 5345 5345
Fax: 00800 7890 7890
URL: http://www.invitrogen.com

In Vitro Systems & Services GmbH, Am
Kalkberg, D-37520 Osterode, Germany.
Tel: +49 55 22 316 250
Fax: +49 55 22 316 211

Jackson ImmunoResearch Laboratories, Inc.,
872 West Baltimore Pike, PO Box 9, West
Grove, PA 19390, USA.
Tel: 001 610 869 4067
Fax: 001 610 869 0171
URL: http://www.jacksonimmuno.com

Kapak Corporation, 5305 Parkdale Drive,
Minneapolis, Minnesota 55416-1681, USA.
Tel: 952-541-0730   Fax: 952-541-0735
URL: http://www.kapak.com

Leitz, Esselte, GmbH & Co KG,
Siemensstraße 64, 70469 Stuttgart,
Germany.
Tel: +49 711 8103-0
Fax: -486
URL: http://www.leiz.com

Ludl Electronic Products Ltd., 171 Brady
Avenue, Hawthorne, NY 10532, USA.
Tel: 888-769-6111   Fax: 001 914 769 4759
URL: http://www ludl.com

Merck Eurolab GmbH, D-64271 Darmstadt,
Germany.
Tel: 0049-6151-723000
Fax: 0049-6151-72333
URL: http://www.merckeurolab.com

Merck Sharp & Dohme, Research
Laboratories, Neuroscience Research Centre,
Terlings Park, Harlow, Essex CM20 2QR, UK.
URL: http://www.msd-nrc.co.uk
MSD Sharp and Dohme GmbH, Lindenplatz
1, D-85540, Haar, Germany.
URL: http://www.msd-deutschland.com

Metasystems, Inc., 6 Corporation Center
Drive, Broadview Heights, Ohio 44147, USA.
Tel: 800-788-8283   Fax: 440-526-1406
URL: http://www.metasystems.com

Midland Certified Reagent Company, 3112-A
West Cuthbert Avenue, Midland, TX 79701,
USA.
Tel: 800-247-8766   Fax: 915-694-2387
URL: http://www.mcrc.com

Millipore (UK) Ltd., The Boulevard,
Blackmoor Lane, Watford, Hertfordshire
WD1 8YW, UK.
Tel: 01923 816375
Fax: 01923 818297
URL: http://www.millipore.com/local/UK.htm
Millipore Corp., 80 Ashby Road, Bedford,
MA 01730, USA.
Tel: 001 800 645 5476
Fax: 001 800 645 5439
URL: http://www.millipore.com

MJ Research, 1250 Elko Drive, Sunnyvale,
CA 94089-2213, USA.
Tel: 800-752-8585   Fax: 408-734-0300
URL: http://www.mjresearch.com

Molecular Probes Inc., PO Box 22010,
Eugene, OR 97402-0469, USA.
Tel: 541-465-8300   Fax: 541-344-6504
URL: http://www.probes.com
Molecular Probes Europe BV, PoortGebouw,
Rijnsburgerweg 10, 2333 AA Leiden, The
Netherlands.
Tel: +31-71-5233378
Fax: +31-71-5233378
URL: http://www.probes.com

Murex Biotech Ltd., Central Road, Temple
Hill, Dartford, Kent DA1 5LR, UK.
Tel: 01322 277711   Fax: 01322 273288
URL: http://murexuk.add.abbott.com/dart-
ford/website3/

**NEN (New England Nuclear) Life Science
Products**, 549 Albany Street, Boston, MA
02118, USA.
Tel: 800-551-2121
Fax: 617-482-1380
URL: http://www.nen.com

New England Biolabs, 32 Tozer Road, Beverley, MA 01915-5510, USA. Tel: 001 978 927 5054

NIH (National Institutes of Health), Clinical Research, National Center for Research Resources, One Rockledge Centre, Suite 6030, 6705 Rockledge Drive, MSC 7965, Bethesda, Maryland 20892-7965, USA. Tel: 301-435-0790   Fax: 301-480-3661 URL: http://www.nih.gov

Nikon Inc., 1300 Walt Whitman Road, Melville, NY 11747-3064, USA. Tel: 001 516 547 4200 Fax: 001 516 547 0299 URL: http://www.nikonusa.com Nikon Corp., Fuji Building, 2-3, 3-chome, Marunouchi, Chiyoda-ku, Tokyo 100, Japan. Tel: 00813 3214 5311 Fax: 00813 3201 5856 URL: http://www.nikon.co.jp/main/index_e.htm

Nunc (Life Technologies/Nalge Nunc Intl), 75 Panorama Creeke Drive, Rochester, NY 14625, USA. Tel: 716-586-8800 URL: http://www.nalgenunc.com

Nycomed Amersham plc, Amersham Place, Little Chalfont, Buckinghamshire HP7 9NA, UK. Tel: 01494 544000   Fax: 01494 542266 URL: http://www.amersham.co.uk Nycomed Amersham, 101 Carnegie Center, Princeton, NJ 08540, USA. Tel: 001 609 514 6000 URL: http://www.amersham.co.uk

O. Kindler GmbH & Co., Ziegelhof Str. 214, 79110 Freiburg, Germany.

Omega Optical, Inc., PO Box 573, 3 Grove Street, Brattleboro, VT 05302, USA. Tel: 802-254-2690   Fax: 802-254-3937 URL: http://www.omegafilters.com

Operon Technologies, Inc., 1000 Atlantic Avenue, Suite 108, Alameda, CA 94501, USA. Tel: 001 688 2248   Fax: 001 510 865-5255 URL: http://www.operon.com

P.A.L.M. Microlaser Technologies AG, 82347 Bernried, Germany. URL: http://http://medizin.li_mt_index/_pa/pa-000036578.html

PerkinElmer, 45 William Street, Wellesley, MA 02481-4078, USA. Tel: 781-237-5100 URL: http://www.perkinelmer.com Perkin Elmer Ltd., Post Office Lane, Beaconsfield, Buckinghamshire HP9 1QA, UK. Tel: 01494 676161 URL: http://www.perkin-elmer.com

Pharmacia Biotech (Biochrom) Ltd., Unit 22, Cambridge Science Park, Milton Road, Cambridge CB4 0FJ, UK. Tel: 01223 423723   Fax: 01223 420164 URL: http://www.biochrom.co.uk Pharmacia and Upjohn Ltd., Davy Avenue, Knowlhill, Milton Keynes, Buckinghamshire MK5 8PH, UK. Tel: 01908 661101   Fax: 01908 690091 URL: http://www.eu.pnu.com

Pharmacia Corporation, 100 Route 206 North, Peapack, NJ 07977, USA. Tel: 888-768-5501 Fax: 908-901-8379 URL: http://www.pharmacia.com

Photometrics Ltd., 2440 E. Britannia Drive, Tucson, AZ 85706, USA. Photometrics Gmbh, Sollner Strasse 61, D-81479 Munich, Germany.

Promega UK Ltd., Delta House, Chilworth Research Centre, Southampton SO16 7NS, UK. Tel: 0800 378994   Fax: 0800 181037 URL: http://www.promega.com

Promega Corp., 2800 Woods Hollow Road, Madison, WI 53711-5399, USA.
Tel: 001 608 274 4330
Fax: 001 608 277 2516
URL: http://www.promega.com

Qiagen UK Ltd., Boundary Court, Gatwick Road, Crawley, West Sussex RH10 2AX, UK.
Tel: 01293 422911
Fax: 01293 422922
URL: http://www.qiagen.com
Qiagen Inc., 28159 Avenue Stanford, Valencia, CA 91355, USA.
Tel: 001 800 426 8157
Fax: 001 800 718 2056
URL: http://www.qiagen.com

Qiagen Genomics, Inc., 1725 220th Street SE, Suite 104, Bothell, WA, USA.
Tel: 425-398-3140
Fax: 425-398-3160
URL: http://www.qiagengenomics.com

Quality Biological, 7581 Lindbergh Drive, Gaithersburg, MD 20879, USA.
Tel: 800-443-3331
Fax: 301-840-5450
URL: http://www.qualitybiological.com

Resource Center of the German Human Genome Project, Heubnerweg 6, 14059, Berlin, Germany.
Tel: +49 30 32639 251
URL: http://www.rzpd.de

Robbins Scientific, 1250 Elko Drive, Sunnyvale, CA 94089-2213, USA.
Tel: 800-752-8585
Fax: 408-734-0300
URL: http://www.robsci.com

Roche Diagnostics Ltd., Bell Lane, Lewes, East Sussex BN7 1LG, UK.
Tel: 01273 484644
Fax: 01273 480266
URL: http://www.roche.com

Roche Diagnostics Corp., 9115 Hague Road, PO Box 50457, Indianapolis, IN 46256, USA.
Tel: 001 317 845 2358
Fax: 001 317 576 2126
URL: http://www.roche.com
Roche Diagnostics GmbH, Sandhoferstrasse 116, 68305 Mannheim, Germany.
Tel: 0049 621 759 4747
Fax: 0049 621 759 2890
URL: http://www.roche.com
Roche Diagnostics, 201 Boulevard Armand Frappier, Laval, Quebec H7V 4A2, Canada.
Tel: 405-686-7050
Fax: 405-686-7011
URL: http://www.roche.com

Rockland Inc., PO Box 316, Gilbertsville, PA 19525, USA.
Tel: 610-369-1008   Fax: 610-367-7825
URL: http://www.rockland-inc.com

Roper Scientific MASD, Inc., 11633 Sorrento Valley Road, San Diego, CA 92121, USA.
Tel: 800-462-4307
URL: http://www.masdkodak.com

Sarstedt Inc., 6373, Des Grandes Prairies, St. Leonard, Quebec H1P 1A5, Canada.
Tel: 514-328-6614   Fax: 514-328-9391
URL: http://www.sarstedt.com

Schleicher and Schuell, 10 Optical Avenue, PO Box 2012, Keene, NH, USA.
Tel: 800-245-4024   Fax: 603-357-7700
URL: http://www.s-and-s.com

SeeDNA Biotech Inc., Metropolitan Bld, 443 Ouellette Avenue, Suite 100, Windsor, Ontario N9A 4J2, Canada.
Tel: 519- 252-8669   Fax: 519-252-2915
URL: http://www.seedna.com

Shandon, 171 Industry Drive, Pittsburgh, PA 15275, USA.
Tel: 800-245-6212   Fax: (412) 747-4080
URL: http://www.shandon.com

Shandon Scientific Ltd., 93-96 Chadwick Road, Astmoor, Runcorn, Cheshire WA7 1PR, UK.
Tel: 01928 566611
URL: http://www.shandon.com

Shandoz Lipshaw, 171 Industry Drive, Bellwood, IL 60104, USA.
Tel: 800-245-6212   Fax: 412-788-1138

Sigma Aldrich Canada Ltd., 2149 Winston Park Dr, Oakville, Ontario L6H 6J8, Canada (Fluka).
Tel: 905-829-9500
URL: http://sigmaaldrich.com

Sigma-Aldrich Co. Ltd., The Old Brickyard, New Road, Gillingham, Dorset XP8 4XT, UK.
Tel: 01747 822211
Fax: 01747 823779
URL: http://www.sigma-aldrich.com
Sigma-Aldrich Co. Ltd., Fancy Road, Poole, Dorset BH12 4QH, UK.
Tel: 01202 722114
Fax: 01202 715460
URL: http://www.sigma-aldrich.com
Sigma Chemical Co., PO Box 14508, St Louis, MO 63178, USA.
Tel: 001 314 771 5765
Fax: 001 314 771 5757
URL: http://www.sigma-aldrich.com

Sorvall, 31 Pecks Lane, Newtown, CT 06470-2337, USA.
Tel: 800-522-7746   Fax: 203-270-2166
URL: http://www.sorvall.com

Southern Biotechnology Assoc. Inc., 160A Oxmoor Boulevard, Birmingham, AL 35209, USA.
Tel: 800-722-2255   Fax: 205-945-8768
URL: http://southernbiotech.com

Stratagene Inc., 11011 North Torrey Pines Road, La Jolla, CA 92037, USA.
Tel: 001 858 535 5400
URL: http://www.stratagene.com

Stratagene Europe, Gebouw California, Hogehilweg 15, 1101 CB Amsterdam Zuidoost, The Netherlands.
Tel: 00800 9100 9100
URL: http://www.stratagene.com

Surgipath Medical Industries Inc., PO Box 528, Richmond, IL 60071, USA.
Tel: 800-225-3035
Fax: 815-678-2216
URL: http://www.surgipath.com

Ted Pella, Inc., PO Box 492477, Redding, CA 96049-2477, USA.
Tel: 530-243-2200
Fax: 530-243-3761
URL: http://www.tedpella.com

TeleChem International, Inc. // ArrayIt.com, 524 E. Weddell Drive, Suite 3, Sunnyvale, CA 94089, USA.
Tel: 408-744-1331
Fax: 408-744-1711
URL: http://www.arrayit.com

TGS Corporate Headquarters, 5330 Carroll Canyon Road, Suite 201, San Diego, CA 92121-3758, USA.
Tel: ++1 858 457-5359 or (800) 544 4847
Fax: ++1 858 452 2547
URL: http://www.tgs.com
TGS European Headquarters, P.A. Kennedy, I-BP 227, Avenue Henri Becquerel, 33708 Mérignac Cedex, France.
Tel: ++33 5 5613 37 77
Fax: ++33 5 5613 02 10
URL: http://www.tgs.com

Thermatron, 687 Lowell Street, Methuen, MA 01844, USA.
Tel: 978-687-8844
Fax: 978-687-2477
URL: http://www.thermatroneng.com

United States Biochemical, PO Box 22400, Cleveland, OH 44122, USA.
Tel: 001 216 464 9277

Universal Imaging Corporation, 502 Brandywine Parkway, West Chester, PA 19380, USA.
Tel: 610-344-9410
Fax: 610-344-9515
URL: http://www.image1.com

VWR Canlab, 2360 Argentia Road, Mississauga, Ontario L5N 5Z7, Canada.
Tel: 800-932-5000
URL: http://www.vwrcanlab.com

Vector Laboratories, Inc., 30 Ingold Road, Burlingame, CA 94010, USA.
Tel: 800-227-6666
Fax: 650-697-0339
URL: http://www.vectorlabs.com

Ventana Medical Supplies, Inc., 3865 N. Business Center Drive, Tuscon, AZ 85705, USA.
Tel: 800-227-2155
Fax: 520-887-2558
URL: http://www.ventanamed.com

Virtis, 815 Route 208, Gardiner, NY 12525, USA.
Tel: 800-765-6198
Fax: 845-255-5338
URL: http://www.virtis.com

Vysis Inc., 3100 Woodcreek Drive, Downers Grove, IL 60515-5400, USA.
Tel: 800-553-7042
Fax: 630-271-7138
URL: http://www.vysis.com

Wheaton, 1101 Wheaton Avenue, Millville, NJ 08332, USA.
Tel: 856-825-1400 ext. 2471
Fax: 856-293-6330
URL: http://www.wheaton.com

Index

adherent cells, cultivation and
        fixation 125
ageing 63–4, 65, 66, 187
Alexa dyes 101
alkaline buffer, chromatin fibre
        preparation 79, 81–2
alkaline denaturation 105
alkaline lysis preparation 9–10
amine-reactive reagents,
        chemical forms 18
aminomethylcoumarin (AMCA)
        107
Angelman syndrome 199
antifade
    fibre FISH 90
    3D FISH 137–8
argon ion laser 6
armFISH 208

bacterial artificial chromosome
        (BAC) DNA preparation
        11–13
banding technologies 1
beam scanning 146
beam splitter 148–9
biotin labelled probe detection
        46–7
bladder cancer 202
blocking slides 231
bovine serum albumin (BSA)
        removal 25
breast cancer 201–2
bromodeoxyuridine (BrdU) 7,
        35–6

cDNA
    bacterial clone cultures 223,
        225

direct labelling with
        fluorescent dyes 234
cDNA microarrays 221–40
    applications 239
    clone cultures 223, 225
    data mining and statistical
        analysis 238–9
    fabrication 223–31
    hybridization 235–6
    image acquisition and
        analysis 237–8
    post-hybridization washes
        236
    sensitivity and specificity 238
    slide printing 230
    target production 232–5
    target purification 234–5
cDNA probes 109–10
cell culture 34–6, 186–7
cell fixation 56, 58–60
    mapping studies 38–9
    3D FISH 123–7
cell preparation
    RNA FISH 96–8
        cytogenetic cells 97–8
        detergent extraction 96–7
        3D FISH 123–7
cell synchronization 34, 35–6
centromere probes 184
chemically synthesized probes
        9
chimerism 13
chromatic shift 150, 151–2
chromatin fibre 78, 79–83
    alkaline buffer 79, 81–2
    cytospin preparation 79,
        82–3
    drug treatment 79, 80–1
    high pH buffer approach 79
    nuclear envelope opening 79

chromatin structure
    fibre FISH studies 78
    preservation during 3D FISH
        144
chromosome painting 72
    hybridization with 109
    probes 15
    RNA FISH and 108–9
chromosomes, counterstaining
        and banding 47–8; see
        also murine chromosome
        preparation
clinical cytogenetics 3, 183–203
    probes 184–5, 198–202
    results assessment and
        reporting 194–8
    sample preparation 185–94
    web sites 202
clone
    culture 223, 225
    PCR amplification 226–7
cloned probes 8–9
    preparation 9–15
colcemid 34–5, 56
comparative genomic
        hybridization (CGH) 3,
        159–82
    advantage 159
    genomic DNA amplification
        by universal PCR 168–71
    genomic DNA isolation
        161–2
        blood 165–6
        paraffin-embedded tissue
            167
        solid tissue 166–7
    image acquisition and
        evaluation 177–8
    in situ hybridization 176–7
    matrix CGH 3, 180–1

251

comparative genomic
      hybridization (*contd.*)
  metaphase chromosomes
      161, 164-5
    denaturation 174-5
  micromanipulation of single
      cells 167-8
  probe labelling 172-4
  probe mixture and
      denaturation 175-6
  single cell (SCOMP) 169,
      170-1
  troubleshooting 178-80
confocal microscopy 113,
      144-54
  beam scanning 146
  beam splitter 148-9
  calibration 150
  chromatic shift 150, 151-2
  distance measurements
      152-3
  filters 147-9
  image acquisition 149-50
  image processing software
      154
  image stack 145
  mounting medium 149
  noise 149-50
  pinhole diameter 149
  projections 152
  range indicator colour maps
      150
  resolution 154
  stage scanning 146
  surface rendering 152
  3D-slicer 152
  visualization 150, 152
  volume measurements 153-4
  volume rendering 152
  voxel size 149
control DNA 162
convolution 154
cooled charged-couple device
      (CCD) camera 48, 114
cosmid DNA preparation
      11-13
Cot-1 DNA 101
counterstaining
  fibre FISH 90
  mapping studies 47-8
  3D FISH 137-8
coverslips 149
  preparation 123-4
Cri du chat 199
cyclicons 17
Cytocell Inc. 184

cytogenetic preparations 97-8,
      189
cytological preparations,
      interphase nuclei from
      189
cytospin, chromatin fibre
      preparation 79, 82-3

DAPI 137
DAPI-banding 47-8
deconvolution 147, 153, 154
degenerate oligonucleotide
      primed PCR; *see* DOP-PCR
deproteinization 128
destaining 187
detergent extraction 96-7
DiGeorge syndrome 199
digital imaging 114-16
digoxigenin (DIG) labelled
      probe detection 46-7
DNA
  alkaline denaturation 105
  coupling detection with RNA
      105-6
  heat denaturation 104
  hybridization 103-5
DNA fibre 78, 83-5
  gel block preparation 84-5
DNase treatment 25
DOP-PCR
  genomic DNA amplification
      169-70
  probe libraries 15-16

electron microscopy 113
enzymatically amplified probes
      9, 15-16
ethanol precipitation 32-3
Ewings sarcoma 202
exon suppression 110-12

fibre FISH 3, 77-92
  applications 77-8
  chromatin fibre 78, 79-83
    alkaline buffer 79, 81-2
    cytospin preparation 79,
      82-3
    drug treatment 79, 80-1
    high pH buffer approach
      79

nuclear envelope opening
      79
  counterstaining and antifade
      90
  DNA fibre 78, 83-5
    gel block preparation
      84-5
  equipment 78-9
  fibre release 78
  fibre type selection 78
  history 77
  hybridization 86-8
  photography 91
  post-hybridization wash 88-9
  probe labelling 85-6
  signal amplification 89-90
  steps 85-90
fibroblast cultures 37
filters 7-8, 48, 113-14
  confocal microscopy 147-9
fixative/fixation 56, 58-60
  mapping studies 38-9
  3D FISH 123-7
fluorescence, principles 5-8
fluorescence energy transfer
      (FRET) 6
fluorescence *in situ*
      hybridization (FISH) 70-1
  advantages 3
  history 2
  success, parameters ensuring
      66-7
  versatility 2
fluorescent bead slides 151
fluors 5, 6-7
  choice 6
  commonly used 7, 8
  coupling to amine-modified
      nucleic acids 26
  coupling to nucleotides
      17-19
  direct chemical coupling 26
  key characteristics 5-6
  storage 18
free chromatin FISH 77

G-banding 63
  destaining 187
gel block preparation 84-5
gene amplification 201-2
genomic DNA
  amplification by PCR 168-71
  isolation 161-2
    blood 165-6

paraffin-embedded tissue
167
solid tissue 166–7
genomic probes 49
Giemsa–trypsin banding 63–5
glioblastoma 202

H₂O₂ slide pre-treatment 64, 65
haematological cancer, 199, 201
haptens 7
coupling to amine-modified
nucleic acids 26
coupling to nucleotides
17–19
detection
multicolour FISH 214
SKY 213–14
3D FISH 136–7
direct chemical coupling 26–7
heat denatured DNA 104
helium neon laser 6
HER2/Neu 201–2
hybridization
cDNA microarrays 235–6
chromosome paint 109
detection following FISH 71–2
to DNA 103–5
efficiency 194–5
fibre FISH 86–8
pre-treatments 128–31
to RNA 100–3
3D FISH 131–4
hypotonic treatment 38–9

icthyosis, X-linked 199
image analysis 214–15
cDNA microarrays 237–8
comparative genomic
hybridization 177–8
mapping studies 48
multicolour FISH 216
SKY 215–16
IMAGE consortium 223
image processing software 154,
178, 237
image stack 145
immunodetection
mapping studies 45–7
and 3D FISH 142–3
immunostaining for proteins
107–8
ink-jetting 180, 222

in situ hybridization 176–7
interphase FISH
clinical cytogenetics 187,
189–94
guidelines 196–7
interpretation 197–8
mapping studies 36–7, 53
scoring criteria 196
intersex transplant monitoring
201
intron probes 109–10
isothermal amplification 16

Kallman syndrome 199

lasers 6
leukaemias 201
light sources 6
lissencephaly 199
loop-mediated isothermal
amplification 16
lymphocyte culture and harvest
34–5

mapping, see single copy gene
mapping
matrix CGH 3, 180–1
mercury arc lamps 6
metaphase chromosomes
clinical cytogenetics 185–7,
188–9
comparative genomic
hybridization 161, 164–5
denaturation 174–5
mapping studies 34–6, 49–50,
51–3
multicolour FISH 210–11
SKY 209–10
methotrexate 35
microarray FISH 4, see also
cDNA microarrays
microdeletion syndromes
198–9
micromanipulation of single
cells 167–8
microscopy 48, 113–14
cell selection for 195–6
see also confocal microscopy
Miller–Dieker syndrome 199
mitotic index 62–3

molecular beacons 17
molecular cytogenetics 2
mouse chromosomes 62,
66–76
see also clinical cytogenetics
monolayer cell extraction 97
mononuclear cell isolation
127–8
multicolour FISH (M-FISH) 4,
51–3, 205, 206, 208
commercial probes 208
hapten detection 214
image acquisition and
analysis 216
metaphase spreads 210–11
post-hybridization wash 214
probe labelling 208
probe precipitation and
hybridization 212–13
troubleshooting 216–19
multiple label techniques
105–12
multiple probes, relational
mapping 51–3
multiprobe device 184
murine chromosome
preparation 55–76
banding and in situ
hybridization 55–63
bone marrow cells 57
Giemsa–trypsin banding
63–5
lymph node cells 60–1
molecular cytogenetics 62,
66–76
plasmacytoma 61
splenic cells 60
thymic cells 60–1
MYCN 201

NaOH denaturation 103–4, 105
neuroblastoma 201, 202
nick translation 20–1, 100,
172–4
normalization constant 238
nuclear envelope opening 79
nucleic acid sequence-based
replication 16

oligonucleotide arrays 222
oligonucleotide hybridization
102–3

oligonucleotide probes 16–17
osteogenesis imperfecta 109, 115

P1 artificial chromosome (PAC)
    DNA preparation 11–13
paraffin-embedded tissue
    genomic DNA isolation 167
    isolation of intact nuclei from 193–4
    tissue section preparation 189–92
PCR, *see* polymerase chain reaction
Pefabloc SC 25
pellet fixation 56, 58
pepsin concentration 67
pepsin digestion 128, 129–30
peptide nucleic acid 17
phage DNA preparation 10–11
photobleaching 6, 7
photography, fibre FISH 91
phycoerythrin 26
plasmid DNA
    isolation 226
    preparation 9–10
point spread function 154
poly-L-lysine slide pre-treatment 229–30
polymerase chain reaction (PCR)
    clone amplification 226–7
    genomic DNA amplification 168–71
    labelling 24–5
    probe amplification 15–16
    product purification 228
    product quantification 227–8
post-fixation treatments 130–1
post-hybridization washes
    cDNA microarrays 236
    fibre FISH 88–9
    mapping studies 44–5
    multicolour FISH 214
    SKY 213–14
    3D FISH 134
post-labelling DNA processing and purification 25
Prader–Willi syndrome 199
pre-labelled reagents 17
probe libraries 15–16
probes 8–17
    clinical cytogenetics 184–5, 198–202

denaturation and hybridization
    fibre FISH 86–8
    mapping studies 42–3
    3D FISH 132–4
detection
    mapping studies 46–7
    3D FISH 135–7
labelling 19–25, 26–7, 66–7, 68–9
    comparative genomic hybridization 172–4
    efficiency 68–9
    fibre FISH 85–6
    multicolour FISH 208
    nick translation 20–1, 100, 172–4
    PCR 24–5
    quality 69–70
    random primer 21–2
    RNA FISH 99
    RNA transcription 22–3
    SKY 208
    3D FISH 131–2
mapping studies 30–3, 49–53
mixture 175–6
PCR amplification 15–16
precipitation and hybridization
    multicolour FISH 212–13
    SKY 211–12
preparation 9–15
    mapping studies 32–3
    RNA FISH 99
    3D FISH 132
purification 66
titration 68–9
types 8–9
web sites 30–2

random primer labelling 21–2
R-banding 47
red blood cells, lysing 56
relational mapping 51–3
repeat sequence probes, denaturation 42
replication labelling 139–42
reverse G-banding 47–8
rhabdomyosarcoma 202
RNA
    coupling detection with DNA 105–6
    extraction 232–3

hybridization 100–3
transcription labelling 22–3
RNA FISH 3, 93–118
    cell preparation 96–8
        cytogenetic preparations 97–8
        detergent extraction 96–7
    chromosome paints and 108–9
    digital imaging 114–16
    hybridization to DNA 103–5
    hybridization to RNA 100–3
    microscopy 113–14
    multiple label techniques 105–12
    probe preparation 99
    protein detection coupled with 107–8
    visualization 112–16

SAGE 221–2
sample preparation 66
    clinical cytogenetics 185–94
sandwich effect 215
self-sustaining sequence replication 16
signal amplification 89–90
single cell comparative genomic hybridization (SCOMP) 169, 170–1
single copy gene mapping 29–54
    chromosome counterstaining and banding 47–8
    image analysis 48
    immunodetection 45–7
    microscopy 48
    post-hybridization washes 44–5
    probes 30–2, 49–53
        denaturation and hybridization 42–3
        preparation 32–3
    slide preparation 39–42
    target DNA
        denaturation and hybridization 42–4
        preparation 34–9
single copy genomic probes 32
    denaturation 42
slides
    ageing 63–4, 65, 66, 187
    blocking 231
    denaturation 67

drying 186
preparation
    clinical cytogenetics 186
    fibre FISH 87
    mapping studies 39–42
    3D FISH 123–4
pre-treatment 41–2
    H$_2$O$_2$ slide 64, 65
    poly-L-lysine 229–30
printing 230
storage 187
spectral karyotyping (SKY) 74,
        205, 206–8
    banding after 76
    commercial probes 206
    hapten detection 213–14
    image acquisition and
        analysis 215–16
    metaphase spreads 209–10
    post-hybridization wash
        213–14
    probe labelling 208
    probe precipitation and
        hybridization 211–12
    troubleshooting 216–19
spectral precision distance
        microscopy (SPDM) 153
spot test 69–70
stage scanning 146
Stokes shift 5
strand displacement
        amplification 16
stringency 44
succinimidyl esters 18
    protein labelling 26–7
suppliers 241–9
surface rendering 152
suspended cells, preparation
        and fixation 126–7

suspension fixation 59–60
synchronized cell cultures 34,
        35–6
synthetic oligonucleotide
        probes 16–17

target DNA
    denaturation and
        hybridization 42–4
    preparation 34–9
target production 232–5
target purification 234–5
Thermotron 39
three-colour fusion probes 199,
        201
three-dimensional (3D) FISH 3,
        119–57
    cell preparation and fixation
        123–7
    chromatin structure
        preservation 144
    counterstaining and antifade
        137–8
    halogenated nucleotide
        detection after 140–2
    hybridization 131–4
    post-hybridization wash 134
    pre-treatments 128–31
    probe
        denaturation and
            hybridization 132–4
        detection 135–7
        labelling 131–2
        preparation 132
    protein immunodetection
        and 142–3

replication labelling and
        139–42
tissue arrays 190–1
transcript differentiation
        109–10
Triton treatment 96
trypsin solution 64
two-photon microscopy 146
tyramide amplification 45

unique sequence probes 32,
        49–50
    denaturation 42

velocardiofacial syndrome 199
volume rendering 152
voxel size 149

web sites 30–2, 202
Williams syndrome 199
Wolf-Hirschhorn syndrome
        199

xenon arc lamps 6
XIST RNA 108

yeast artificial chromosome
        (YAC) DNA preparation
        13–15